Advances in LUMINESCENCE SPECTROSCOPY

Cline Love / Eastwood, editors

ASTM STP 863

ADVANCES IN LUMINESCENCE SPECTROSCOPY

A symposium sponsored by
ASTM Committee E-13 on
Molecular Spectroscopy
Atlantic City, NJ, 7 March 1983

ASTM SPECIAL TECHNICAL PUBLICATION 863
L. J. Cline Love, Seton Hall University, and
DeLyle Eastwood, U.S. Army Corps of
Engineers, editors

ASTM Publication Code Number (PCN)
04-863000-39

 1916 Race Street, Philadelphia, PA 19103

Library of Congress Cataloging in Publication Data

Advances in luminescence spectroscopy.

(ASTM special technical publication; 863)
"ASTM Publication code number (PCN) 04-863000-39"
Includes bibliographies and index.
1. Luminescence spectroscopy—Congresses.   I. Cline Love, L. J.   II. Eastwood, D.   III. ASTM Committee E-13 on Molecular Spectroscopy.   IV. Series.

QC476.6.A38   1985   543′.0858   84-71320
ISBN 0-8031-0412-X

Copyright © by AMERICAN SOCIETY FOR TESTING AND MATERIALS 1985
Library of Congress Catalog Card Number: 84-71320

NOTE

The Society is not responsible, as a body,
for the statements and opinions
advanced in this publication.

Printed in Baltimore, Md.
May 1985

Dedication

With great sorrow we note the death of Professor Gorden F. Kirkbright (1938 to 1984) of the University of Manchester Institute of Science and Technology. After receiving his Ph.D. from the University of Birmingham, he did postdoctoral studies in the United States and Austria and then joined the staff of Imperial College in London in 1964. During his years there, he did extensive research in atomic absorption and fluorescence, molecular fluorescence and phosphorescence, photoacoustic, and atomic emission spectrometry with emphasis in the latter on inductively coupled plasmas. During this time, he retained an active interest in the affairs of a London-based company, EDT Research, of which he was a founding director.

In 1980, the University of Manchester Institute of Science and Technology undertook a major step in the development of analytical methodology by forming a new Department of Instrumentation and Analytical Science. Professor Kirkbright was appointed as the first head of the newly established chair in analytical science. The center is recognized today as being a major development in international analytical chemistry, coupling important advances in chemistry with improvements in instrumentation. Professor Kirkbright received the 1983 Gold Medal of the Analytical Division, Royal Society of Chemistry, for his outstanding research work. His list of publications is in excess of 200 papers, and he was on editorial advisory boards of major journals such as The Analysts, ARAAS, Canadian Journal of Spectroscopy, and Spectrochimica Acta. He had recently been appointed Editor-for-Europe for Applied Spectroscopy.

Professor Kirkbright was noted not only for his science, but for his wit and joy of life. Energetic both at work and away, he was a regular figure at major international conferences and at American meetings such as EAS, FACSS, and the Pittsburgh Conference. His scientific contributions and careful scholarship inspired his research students and colleagues alike, and his dedicated classroom teaching profoundly influenced the lives of many young scientists. His creative innovations in several areas of analytical *chemistry should serve as scholarly bench marks to many scientists, both now and in the future. His untimely death at the age of 46 is a shock and profound loss to us all. One of his last scholarly works on "Some Photoacoustic Studies of Consequence in Luminescence Spectroscopy" is included in this publication. This volume is dedicated to Professor Kirkbright.*

Foreword

The symposium on Advances in Luminescence Spectroscopy was held in Atlantic City, NJ on 7 March 1983. ASTM Committee E-13 on Molecular Spectroscopy and its Subcommittee E13.06 on Molecular Luminescence sponsored the meeting. L. J. Cline Love, Seton Hall University, and DeLyle Eastwood, U.S. Army Corps of Engineers, served as chairmen of the symposium and are editors of the publication.

Related ASTM Publications

New Directions in Molecular Luminescence, STP 822 (1983), 04-822000-39

Compilation of Methods for Emission Spectrochemical Analysis, Seventh Edition, 1982, 03-502082-39

ASTM Manual on Practices in Molecular Spectroscopy (E-13), Fourth Edition, 1980, 03-513079-39

An Introduction to X-Ray Spectroscopy, published in Great Britain, 13-109000-39

Quantitative Mass Spectroscopy, published in Great Britain, 13-114000-39

A Note of Appreciation to Reviewers

The quality of the papers that appear in this publication reflects not only the obvious efforts of the authors but also the unheralded, though essential, work of the reviewers. On behalf of ASTM we acknowledge with appreciation their dedication to high professional standards and their sacrifice of time and effort.

ASTM Committee on Publications

ASTM Editorial Staff

Janet R. Schroeder
Kathleen A. Greene
Helen M. Hoersch
Helen P. Mahy
Allan S. Kleinberg
Susan L. Gebremedhin

Contents

Introduction   1

PROBES OF THE CHEMICAL MICROENVIRONMENT

Spectroscopic Probing of Solvation Interactions — MARY J. WIRTH, AARON C. KOSKELO, CAROL E. MOHLER, DAVID A. HAHN, AND MATTHEW J. SANDERS   5

Analytical Applications of Proton-Transfer Spectroscopy. Detection of Trace Hydrogen-Bonding Impurities in Hydrocarbon Solvents Using 3-Hydroxyflavone as a Fluorescence Probe — DALE MCMORROW AND MICHAEL KASHA   16

Global and Nonglobal Rotations in Proteins Detected by Fluorescence Polarization — RAYMOND F. CHEN AND CARRIE H. SCOTT   26

Comparison of Techniques for Generating Room Temperature Phosphorescence in Fluid Solution — ROBERT WEINBERGER, KAREN REMBISH, AND L. J. CLINE LOVE   40

COUPLED PHENOMENA IN LUMINESCENCE

Some Photoacoustic Studies of Consequence in Luminescence Spectroscopy — GORDON F. KIRKBRIGHT   55

Metal Ion Sensors Based on Immobilized Fluorogenic Ligands — W. RUDOLF SEITZ, LINDA A. SAARI, ZHANG ZHUJUN, STEVEN POKORNICKI, ROBERT D. HUDSON, STEVEN C. SIEBER, AND MAURI A. DITZLER   63

MANIPULATION OF LUMINESCENCE OBSERVABLES

Synchronous-Excitation Fluorescence Applied to Characterization of Phenolic Species — FRANCIS J. PURCELL, RAYMOND KAMINSKI, AND RALPH H. OBENAUF   81

A Concise Feature Set for the Pattern Recognition of Low-Temperature Luminescence Spectra of Hazardous Chemicals — GENE SOGLIERO, DELYLE EASTWOOD, AND JAMES GILBERT 95

Summary 117

Index 121

Introduction

This special technical publication, resulting from the ASTM sponsored symposium on Advances in Luminescence Spectroscopy held at the 1983 Pittsburgh Conference on Analytical Chemistry and Applied Spectroscopy, represents an ongoing effort by the ASTM to acquaint the scientific community with some of the most recent advances in fluorescence and phosphorescence research. It represents, together with its companion publication, *New Directions in Molecular Luminescence* (*STP 822*, 1983), an overview of the major research areas challenging the present day spectroscopist. The scope is diverse, with coverage of new instrumentation, studies of fundamental photophysical phenomena, and analytical applications. The surge in luminescence research in recent years, along with the coupling of luminescence measurements with other techniques, such as chromatography and antibody/antigen reactions, requires a continuous dissemination of recent fundamental discoveries and uses of luminescence spectroscopy. Part of the responsibilities of the ASTM Subcommittee E13.06 on Molecular Luminescence is to facilitate access to these latest advances in analytical, physical, and biochemistry relating to fluorescence/phosphorescence. The functions of the E13.06 Subcommittee, as part of Committee E-13 on Molecular Spectroscopy, extends beyond this to formulating standard methods and practices, reference standards, definitions, and conducting round-robin testing. The work of the committee is done by committed, volunteer experts working in the area of molecular luminescence in institutions throughout the world, and it is hoped that this volume will attract many of its readers to join the subcommittee in these efforts.

The 1983 symposium was designed to complement the first one held in 1982 by including new topics and speakers whose work had not been published in the first volume. A large number of research areas had not been included in the first symposium because of space/time constraints, and the present volume attempts to expand coverage to areas not included previously, as well as include new areas that have emerged since 1982. One general area covered here by four papers is microenvironmental effects on luminescence properties. The often subtle effects of the immediate chemical environment of lumiphors can produce dramatic changes in the spectroscopic observables, as illustrated by studies of solvation interactions, rapid intramolecular proton transfer sensitive to hydrogen-bonding impurities, rotations in proteins, and induction of phosphorescence at ambient temperature. The second area concerns more indirect probes of luminescence using photoacoustic spectroscopy and immobilized fluorescent ligands. A third area covered involves transformation of hazardous chemical luminescence spectral data by synchronous excitation fluorescence and pattern recognition. Taken as a body, these papers reveal the rich diversity of research interests and applications in luminescence spectroscopy, collected for the reader by the ASTM into one convenient source.

Much work remains to be done to meet the creative challenges in the development of a better understanding of the interaction of light with molecules. Scientists, working individually in industry, government, and academia and collectively at the ASTM, are diligently seeking a better understanding of these interactions and ways to translate this understanding for practical use, aided by new computer-assisted instrumentation, theories, and methodologies. The ASTM Subcommittee E13.06 on Molecular Luminescence stands ready to contribute to these efforts by providing a common language and forum to help merge this diverse body of knowledge into a convenient, focused form.

In dedicating this volume to the memory of Prof. Gordon F. Kirkbright, we pay tribute to the scholar and to the man who was a generous, concerned friend to both of us. We join in expressing our sympathies to his wife Ann and two children, Suzanne and Clare.

Acknowledgment

The generous support of major fluorescence instrumentation companies (Farrand, Perkin Elmer, SLM and SPEX) provided a congenial atmosphere for the authors, associates and attendees at the symposium. Special thanks go to SPEX Industries, Metuchen, New Jersey, for hosting a luncheon for the speakers the day of the symposium, and to Thomas J. Porro, Perkin-Elmer Corp., for arranging a reception for participants and colleagues afterwards. The editors also thank the diligent help of the staff at the ASTM. We are indebted to Peter Keliher for contributing the Dedication to Prof. Kirkbright.

L. J. Cline Love
Seton Hall University, Chemistry Department
South Orange, NJ 07079; symposium
co-chairman and co-editor.

DeLyle Eastwood
U.S. Army Corps of Engineers
Missouri River Division Laboratory
420 S. 18th St., Omaha NF 68102;
symposium co-chairman and co-editor.

Probes of the Chemical Microenvironment

Mary J. Wirth,[1] *Aaron C. Koskelo,*[1] *Carol E. Mohler,*[1] *David A. Hahn,*[1] *and Matthew J. Sanders*[1]

Spectroscopic Probing of Solvation Interactions

REFERENCE: Wirth, M. J., Koskelo, A. C., Mohler, C. E., Hahn, D. A., and Sanders, M. J., "**Spectroscopic Probing of Solvation Interactions,**" *Advances in Luminescence Spectroscopy, ASTM STP 863,* L. J. Cline Love and D. Eastwood, Eds. American Society for Testing and Materials, Philadelphia, 1985, pp. 5–15.

ABSTRACT: Three spectroscopic methods have been developed to probe structures of solvation environments in organic solutions: electronic band broadening measurements, two-photon perturbation studies, and fluorescence depolarization experiments. The results of each spectroscopic study indicate that the method does sense solvation structural effects rather than a completely amorphous liquid. The application of these probes is especially promising for investigations of liquid chromatographic retention mechanisms, and a preliminary study of the separation of aromatic isomers is discussed.

KEY WORDS: spectroscopy, solvation, fluorescence, two-photon, rotational diffusion, bandwidth

An understanding of solvation phenomena is important to a wide range of chemical problems. In analytical chemistry, solvation processes affect spectroscopic properties dramatically and are also the primary controlling factors in chromatographic separations. While gas phase and solid state physics are well developed theoretically, similar development of the liquid state has emerged much more recently. Liquids are too amorphous and dynamic to be amenable to solid state theory yet lack the random position correlations of gases that would allow a simple statistical approach. Previously, theoretical descriptions of liquid state phenomena relied upon the assumption that the liquid is an isotropic and continuous medium. This, of course, is an inaccurate picture on the molecular scale, but it offers a reasonable approximation since the large variety of structures provide an average that is nearly amorphous. The availability of large-scale computation has accelerated a fundamentally different tact for liquid state theory, that is, the liquid of interest can be simulated on the molecular scale to allow the

[1]Assistant professor and graduate research assistants, respectively, Department of Chemistry, University of Wisconsin—Madison, 1101 University Ave., Madison, WI 53711.

origins of the macroscopic properties to be studied. The new theoretical approach is particularly valuable to analytical chemistry, where the unique structure of each type of molecule needs to be considered in separating, identifying, or selectively quantifying species.

Perhaps the most important area of analytical chemistry relying on unique properties of molecules in liquids is chromatography. The basis for separating any two compounds is that they have a different change in chemical potential between two solvents. One would wish that liquid theory were sufficient to allow prediction of the chemical potential from the solute and solvent structures. In principle, Monte Carlo simulations allow the direct computation of the chemical potential, provided that the intermolecular interactions can be expressed accurately and that the computer size and time limitations are generous. In practice, the method is applied to simple systems, such as a neat liquid consisting of hard spheres. The organic solutes, mixed solvents, and bonded stationary phases used routinely in chromatography have too complex of an array of intermolecular interactions for present techniques. The gap between theory and practice is thus a cavernous one.

A bridging of the gap between understanding and using the complex materials of modern chromatography can be begun by using spectroscopic measurements to isolate specific types of intermolecular interactions that are pertinent to the separation, for example, dipolar, dispersive, or repulsive interactions. Such measurements would test whether or not a type of interaction were important to the separation, thus providing very useful information for design improvements. Because of the many interactions and the large distribution of structures, this is a very challenging problem. Our group has been investigating several types of spectroscopic measurements for their value in probing chromatographic phenomena. Three techniques are discussed in this paper: band broadening, two-photon spectroscopy, and rotational diffusion measurements. Each senses molecular structural effects in a different way.

Distributions of Solvation Structures

The particular emphasis of our work is in understanding the role of molecular shape on solvation interactions. Shape is important because chromatographic separation of compounds differing only slightly in shape, such as isomers, is especially difficult and would thus benefit from improved understanding. A study of molecular shape effects inherently involves consideration of the structure of the solvation environment.

There is no well defined solvation structure in the same sense as the structure of a molecule or a crystal. One has to abandon the notion that structures, such as those of metal ion complexes or crystal lattices, will be observed because for a liquid to flow large displacements in positions of molecules must be allowed. The sense in which there is structure in liquids is that spatial correlations between solute and solvating molecules occur because repulsive or "steric" interactions restrict the number of configurations. For example, the carbon backbone of

n-hexane cannot approach benzene closely in the benzene plane because the hydrogens would strongly repulse one another. But close approach can occur out of the molecular plane, allowing larger attractive interactions. The latter "structure" is thus expected to predominate over the former. Spectroscopic study of solvation structures involves discerning predominant position correlations and, in the application of interest, evaluating their contributions to the chromatographic separation.

Progress has been made in the study of structures of pure liquids through the use of tools such as neutron diffraction measurements, Raman and infrared spectroscopy, and computer simulations [1]. Direct information about the average number of nearest neighbors is obtainable from diffraction studies, and less direct information about the nature and distribution of interaction is obtainable from shifts and widths of vibrational bands. These experimental measurements are frequently compared with computer simulations, if the liquid is sufficiently simple. The acquisition of more detailed structural information, such as spatial arrangement of the interacting molecules, remains an experimental challenge.

Compared to liquid studies, solvation studies have the additional experimental constraint that only the molecules immediately surrounding the solute are of interest, thus the techniques must be sensitive and selective. Diffraction techniques are not strongly selective and therefore are not as generally useful as the spectroscopic methods. Some selectivity is affordable in diffraction measurements by using the fact that there is a difference in diffraction efficiency for different nuclei [2], and this has been applied to studying hydration numbers of metal ions. Spectroscopy is a more attractive candidate for the study of solvation interactions because solute chromophores can readily be selected even in very dilute solution. However, neither conventional spectroscopy nor diffraction provides information about the spatial arrangements of solvent molecules. Fortunately, these additional degrees of freedom are available from nonlinear spectroscopy, particularly two-photon spectroscopy. In the next sections, the physical basis for spectroscopic studies of solvation structures by two-photon measurements is discussed.

Effect of Intermolecular Interactions on Spectroscopic Properties

Before the liquid itself can be studied, one must consider how intermolecular interactions are observed and interpreted from spectroscopic measurements. The interaction of solute with solvent is coulombic, where the outer electrons of the solute interact with those of the solvent through dispersive, inductive, and dipolar forces, and repulsively through collisions. The same electrons involved in the solvation interaction are also involved in spectroscopic transitions. This is the origin of the utility of spectroscopy in probing the solvation interaction.

Figure 1 illustrates the effect of a solvation environment on the energies of the electronic states of the solute. Both long-range attractive interactions and short-range repulsive interactions shift the energies of the states. Since the spatial

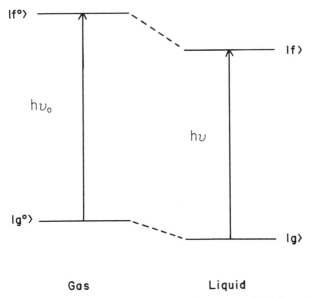

FIG. 1—*Comparison of the ground state and first excited state energy levels for a typical aromatic solute in the gas phase and in an organic liquid. This is to illustrate that each electronic state of a solute interacts differently with the environment.*

distribution of the electron cloud for the excited state is farther from the positive charge centers than for the ground state, the excited state interacts more strongly with the solvent than does the ground state. Very typically, the net interaction is attractive. The result, as illustrated by Fig. 1, is that the transition energy is red shifted. For naphthalene the magnitude of the red shift from the vapor phase to cyclohexane is about 100 cm^{-1}. The theoretical relations between the magnitudes of the solvent-solute interactions and the spectral shifts have been described by Amos and Burrows [3].

An interesting application by Lochmüller et al [4] of solvent shifts to chromatographic phenomena involved the study of the nonuniformity of bonded phases in reverse phase chromatography. Probe molecules were attached to silica gel surfaces that were subsequently derivatized to serve as nonpolar stationary phases. The spectroscopic studies revealed that the derivatization process does not result in uniform surface coverage, as evidenced by significant interactions of the probe molecules with the polar mobile phase. This result is important both to the understanding of observed retention behavior and to column design considerations.

A different type of application of solvent shifts to chromatographic studies, which is under study in our laboratory, involves probing the dynamics of the solvation interaction. One can envision from Fig. 1 that when the strength of the solvation interaction changes in time because of fluctuations in positions of the solvent molecules, then the resonance frequency of the spectroscopic transition changes. The time that an average solute molecule spends at a particular resonance is termed the "coherence lifetime" or "dephasing time." The most rapid fluctuations give rise to broadening that is Lorentzian and is termed "homogeneous" broadening [5]. By the uncertainty principle, the homogeneous width of the spectroscopic band is inversely related to the dephasing time of the transition; thus either time domain of the frequency domain data can be used to measure the fluctuations in the solvation interactions. For an inhomogeneously broadened band, the profile is described by a Voigt function, and the Lorentzian component is still the inverse of the dephasing time. The homogeneous width, or Lorentzian component, is believed to be primarily contributed by repulsive interactions between molecules [6].

Dynamics of the repulsive interactions are of interest in chromatographic studies because these interactions are related to solvation structure. To a first approximation, the more compatible the solute and solvent molecular shapes, the less the amount of repulsive interaction. A separation based upon molecular shape would thus be expected to be spectroscopically detectable. From retention measurements it is known that isomers of polycyclic aromatic hydrocarbons (PAH) apparently are separated based upon shape in reverse phase chromatography [7]. Electronic spectroscopic bandwidth measurements have been applied to a series of dimethylnaphthalene isomers, and it was found that the spectroscopic behavior correlated with the chromatographic retention behavior [8]. A linear relation between the spectroscopic and chromatographic parameters was observed for the nonpolar phase, indicating that the stationary phase contributes to the shape selectivity. This result demonstrates the ability of spectroscopy to probe the very interactions causing a separation and also underlines the role of the nonpolar stationary phase, which had been widely believed to be inert.

Solvation structures cannot be studied directly by the spectroscopic process in Fig. 1 because orientation information is completely absent from the conventional spectroscopic transition indicated. While the transition itself may be polarized, the fact that the solute molecules are randomly oriented precludes acquisition of orientation information. There have previously been no spectroscopic measurements that directly sense spatial arrangements of solvating molecules in liquids. A way that such spatial information can be obtained is to use more than one photon simultaneously interacting with the solute. The relative polarizations of the two photons identify the spatial properties of the solvation interactions.

Figure 2 illustrates two types of transitions involving two polarized photons, representing experiments that this group has explored: (1) measurement of the solvent symmetry perturbations, which we have related to the spatial arrangement of the nearest neighbor solvent molecules and (2) measurement of the rotational diffusion behavior of the solute, which we have shown to be influenced by the structure of the solvation environment. The polarization relations between the two photons are the parameters of interest in both cases. These two experiments are discussed in more detail in the next sections.

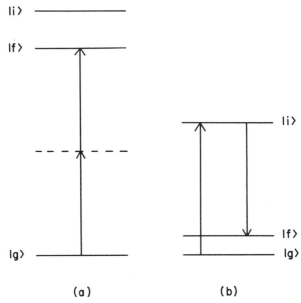

FIG. 2—*Energy level diagrams for* (a) *a simultaneous two-photon transition and* (b) *a fluorescence depolarization experiment. In both cases the relative polarizations of the two photons carry information about the interaction of the solute with its environment.*

Application of Two-Photon Experiments to Probe Solvation Structure

The physical origin of the structural information in two-photon measurements is that the solvation environment perturbs the electronic symmetry of the solute according to the spatial contour of the potential energy in the solvation interaction. The details of the two-photon method for solvation structural studies have been developed using first-order perturbation theory [9]. In this description, new states $|n^0\rangle$ are mixed into a zero order state, such as the final state $|f^0\rangle$, by the Hamiltonian interaction H^1

$$|f\rangle = |f^0\rangle + \sum_{n^0} \frac{\langle n^0|H^1|f^0\rangle}{E_{f^0} - E_{n^0}} |n^0\rangle \qquad (1)$$

The perturbation term is typically about 1% of the zero order term, and the perturbed final state thus has mostly the character of $|f^0\rangle$ and some of the character of $|n^0\rangle$. The presence of the $|n^0\rangle$ character slightly changes the polarization ratio of the two-photon transition according to the magnitude of the perturbation and the symmetry of $|n^0\rangle$.

The characterization of the spatial properties of H^1 is the goal of two-photon experiments because H^1 is determined by the spatial arrangement of the solvent molecules. For the integral $\langle n^0|H^1|f^0\rangle$ to be nonzero, there is a unique relationship between the symmetries of $|n^0\rangle$, $|f^0\rangle$, and H^1. The symmetry properties of H^1 are revealed by virtue of the fact that the two-photon experiment identifies the symmetries of $|n^0\rangle$ and $|f^0\rangle$. Solvation structures can have several components of symmetry, thus the symmetry of H^1 is expressed as a linear combination of the solute point group symmetry representations. For example, hydrogen bonding to one ring of naphthalene would have a large component of long axis antisymmetry, no component of short axis or diagonal anti-symmetry, and a nonzero component of total symmetry. The total symmetry component would disappear if there were an equal but repulsive interaction to the other naphthalene ring.

Experimentally, the characterization of H^1 symmetries is under active investigation. To achieve this goal, calibration of the quantitative amount of mixing of states that occurs upon a given solvation interaction is the next important step in the development of the two-photon perturbation technique. There are two possible approaches to calibration. The first is the ab initio approach where the identity of the mixing states and the magnitudes of the integrals $\langle n^0|H^1|f^0\rangle$ are calculated. Unfortunately, insufficient information prevents reliable results from this approach. The second is the empirical approach where the solute is placed into a variety of known environments, such as crystals, where a single environment can be well characterized, or a highly amorphous fluid, where all symmetry components can be present simultaneously by amounts calculated from statistics. The empirical calibration is presently being developed by this research group for crystal environments as well as for liquid and dense gas environments.

Published results [9] suggest that there is structure present in the solvation environments of aromatic molecules in the solvent methanol. Calibration was accomplished by studying amorphous liquids where only the strength of interaction controls the magnitude of the symmetry perturbation. For methanol, the strength of interaction alone is not sufficient to account for the data, thus structure must be present. A more detailed study of the methanol structure is presently being completed for publication, and the results indicate that the solvent interacts through hydrogen bonding.

A present limitation on the ultimate quantitative interpretation of these experiments is that there is always a distribution of structures contributing to the measured two-photon polarization ratio. Thus, a hypothetical structure having a known polarization ratio cannot be quantitatively compared with a measured solution value because there are also contributions from other structures. It is possible that molecular dynamics simulations of liquids can be useful for extracting quantitative information from the distribution. The reason that a single solvation environment cannot be selected is that is is not possible to obtain the infinitely high wavelength resolution needed to define uniquely one type of solvation interaction because of the fact that the bands are homogeneously broadened by the molecular motions.

The ability to obtain qualitative information on predominant structures is not only an advance in itself but is useful for unraveling chromatographic retention mechanisms. For example, chromatographic structural selection based upon maximum length of the molecule could be distinguished from selection based upon overall size by the type of symmetry perturbations caused by the solvation environment. The maximum length selectivity may be characterized by the average tendency for the solvent molecules, such as n-alkanes, to be aligned with the long solute axis.

The symmetry reasoning used to interpret two-photon behavior can be applied also to Raman spectroscopy. A potential advantage of Raman over two-photon spectroscopy is that the bands are much narrower, eliminating the sometimes severe overlap of vibronic states. The narrower bands of Raman spectroscopy appear because the electronic state is unchanged by the transition, thus the different responses of the ground and excited electronic states do not contribute width of the band. The vibrational ground and excited states respond similarly to one another, thus the broadening is less. The narrower bands in vibrational spectroscopy correspond to longer times over which a transition takes place. Solutes can thus rotate significantly during the Raman transition, diminishing the utility of the polarization data. This loss of polarization information is the major reason that Raman spectroscopy has not yet been implemented for these studies. Experiments involving Raman polarization measurements coupled with rotational diffusion measurements are currently in the design stage. Rotational diffusion measurements also contain solvation information in their own right, and these are discussed in the following section.

Rotational Diffusion Studies

Molecules undergo strongly hindered rotation in liquid solutions. Rotational diffusion for moderately sized organic solutes is described by frictional models where the surrounding molecules strongly damp the random motions of the solute. The nature of the frictional forces are the interactions of the rotating molecule with its environment, particularly the repulsive interactions that limit the positions in space that the solute molecule can occupy. Thus, instead of free rotation that could be observed by microwave spectroscopy, one observes much slower rotation and very broad rotational energy levels. The Debye model of rotational diffusion [10] relates the friction coefficient of a spherical solute to the size of the sphere and the viscosity and temperature of the liquid. Solute shape has been accounted for by Perrin [11], who derived expressions for the three rotational diffusion constants of a general ellipsoid. The Debye and Perrin results are widely used to describe the rotational diffusion behavior of aromatic solutes, thus the solvent is assumed to be a viscous continuum.

Because it is the liquid environment that controls the rotational motion of the solute, one might expect that the presence of solvation structure would influence the relative values of the diffusion constants of the solute. If this were the case, then the relative values of these constants would be solvent dependent rather than determined solely by the shape of the solute.

We have recently undertaken a study designed to detect an influence of solvation structure on the relative values of the diffusion constants. The aromatic molecule tetracene was chosen as the solute because its shape is highly anisotropic, thus its three diffusion constants should be significantly different from one

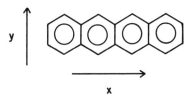

FIG. 3—*The structure and labeling of axes for tetracene. In the fluorescence depolarization experiment, tetracene was excited into its third excited singlet state, having x polarization, and fluorescence was emitted from the first excited singlet state, having y polarization.*

another. Also, the spectroscopic transitions of tetracene are polarized along the same axes as its principle moments of rotational diffusion, thus simplifying the interpretation of the experimental data. Tetracene and the labeling of its axes are shown in Fig. 3. Tetracene fluoresces from its lowest excited singlet state, which is y-axis polarized. If a fluorescence depolarization experiment is performed, the anisotropy is defined as $R(t)$

$$R(t) = (I_\parallel - I_\perp)/(I_\parallel + 2I_\perp) \qquad (2)$$

where $I_\parallel$ is the fluorescence intensity polarized parallel to the excitation and $I_\perp$ is that polarized perpendicular to the excitation. The functionality of $R(t)$ has been derived by Chuang and Eisenthal for the general asymmetric rotor [*12*]. For the sake of clarity only, $R(t)$ can be simplified by assuming that two of the dynamic axes of tetracene are equal to one another. This equality is not strictly true and the argument can be made without resorting to it; however, it does make the complicated behavior easier to see. By tuning the laser wavelength, the x axis was chosen as the spectroscopic excitation axis. If the x axis is the most rapid axis of rotational diffusion, as the Perrin relations predict, and if the y and z axes are approximately equal, $R(t)$ becomes

$$R(t) = -\tfrac{1}{5} \exp(-6D_z t) \qquad (3)$$

where D_z is the rotational diffusion constant for reorientation about the z axis. The x axis is expected to be the most rapid axis of diffusion because of tetracene's oblong shape. If, however, the z axis is the most rapid axis of rotational diffusion then $R(t)$ has a different functionality

$$R(t) = -\tfrac{3}{10} \exp[-(2D_x + 4D_z)t] + \tfrac{1}{10} \exp(-6D_z t) \qquad (4)$$

where D_x is the constant for rotational diffusion about the x axis.

Equation 3 shows that the anisotropy is always negative when the x axis is the fastest axis of rotational diffusion. Equation 4 offers the interesting possibility that the sign of the anisotropy can change when the z axis becomes the fastest axis of rotational diffusion. Thus, the task of identifying whether or not the relative values of the diffusion constants have changed is made simpler by the need only to determine experimentally whether the sign of the anisotropy is solvent dependent. The possibility of observing a change in sign of the anisotropy occurs only by selection of the excitation axis to be perpendicular to the emission axis.

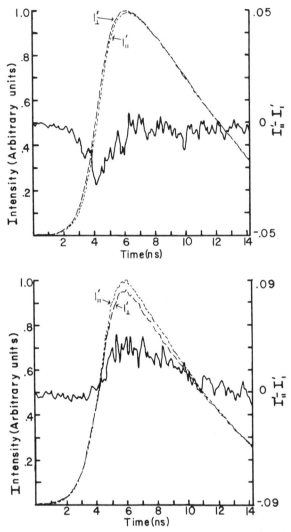

FIG. 4—*Fluorescence depolarization data for tetracene in* n-*dodecanol* (top) *and in ethylene glycol* (bottom) *by time resolution with a fast photomultiplier. The difference curve*, $I_\parallel - I_\perp$, *is calculated from the raw data without deconvolution.*

If they are parallel, then the sign is always positive. Since picosecond pulses have not been conveniently available at blue wavelengths, the experiment has been done by nanosecond technology. Fortunately, the need to observe only the sign of the anisotropy makes the experiment possible by this method.

The fluorescence depolarization behavior of tetracene under x-axis excitation is shown in Fig. 4 for n-dodecanol and ethylene glycol [13]. The data clearly show that the sign of the anisotropy changes thus revealing a change in the relative values of the diffusion constants. We have performed simulations to show that the change must be at least 200% to explain the observed data. In n-dodecanol, tetracene behaves as one would predict from its shape; however, in ethylene glycol tetracene reorients inordinately fast in its molecular plane. The strongly hydrogen bonded structure of the solvent may have a short-range order that has a lower density of solvent in the molecular plane. Other types of solvation studies can help determine the nature of the ethylene glycol solvation structure. The experimental results thus show that solvation structure can dramatically affect the rotational diffusion behavior of solutes.

The tetracene result also suggests that rotational diffusion behavior may be a valuable probe for additional information about solvation structure. For this application, a more general theory must be developed to account for the structure and dynamics of the solvation environment. The rotational diffusion experiment provides complementary information to that of the two-photon and dephasing studies; long-lived solvation structures contribute to rotational diffusion behavior whereas the other two methods probe a "frozen" solvation structure on a short time scale. The rotational diffusion experiments may thus be especially suited to the study of hydrophobic interactions.

Acknowledgment

We are thankful for the financial support by the Graduate School of the University of Wisconsin-Madison and the National Science Foundation. M. J. Sanders acknowledges fellowship support by Eastman Kodak, and D. A. Hahn is grateful for fellowship support from Amoco and from the American Chemical Society, Analytical Division.

References

[1] Murrell, J. N. and Boucher, E. A., *Properties of Liquids and Solutions,* Wiley, New York, 1982.
[2] Narten, A. H. and Hahn, R. L., *Journal of Physical Chemistry,* Vol. 87, 1983, p. 3193.
[3] Amos, A. T. and Burrows, B. T., *Advances in Quantum Chemistry,* Vol. 7, 1973, p. 289.
[4] Lochmuller, C. H., Marshall, D. B., and Wilder, D. R., *Annales De Chimie Acta,* Vol. 30, 1980, p. 130; and Lochmuller, C. H., Marshall, D. B., and Harris, J. M., Vol. 131, 1981, p. 263.
[5] Kubo, R., *Fluctuation, Relaxation and Resonance in Magnetic Systems,* D. Ter Haar, Ed., Plenum Press, New York, 1962.
[6] Schweizer, K. S. and Chandler, D., *Journal of Chemical Physics,* Vol. 73, 1982, p. 5573.
[7] Wise, S. A., Bonnett, W. J., Guenther, F. R., and May, W. E., *Chromatography Science,* Vol. 19, 1981, p. 457.
[8] Wirth, M. J., Hahn, D. A., and Holland, R. A., *Analytical Chemistry,* Vol. 55, 1983, p. 787; and Hahn, D. A. and Wirth, M. J., to be published.
[9] Wirth, M. J., Koskelo, A. C., and Mohler, C. E., *Journal of Physical Chemistry,* Vol. 87, 1983, p. 4397.
[10] Debye, P., *Polar Molecules,* Chemical Catalog Company, New York, 1929.
[11] Perrin, F., *Journal of Physical Radium,* Vol. 5, 1934, p. 497.
[12] Chuang, T. J. and Eisenthal, K. B., *Journal of Chemical Physics,* Vol. 57, 1972, p. 5094.
[13] Sanders, M. J. and Wirth, M. J., *Chemical Physics Letters,* Vol. 101, 1983, p. 361.

Dale McMorrow[1] *and Michael Kasha*[1]

Analytical Applications of Proton-Transfer Spectroscopy. Detection of Trace Hydrogen-Bonding Impurities in Hydrocarbon Solvents Using 3-Hydroxyflavone as a Fluorescence Probe

REFERENCE: McMorrow, D. and Kasha, M., "**Analytical Applications of Proton-Transfer Spectrocopy. Detection of Trace Hydrogen-Bonding Impurities in Hydrocarbon Solvents Using 3-Hydroxyflavone as a Fluorescence Probe,**" *Advances in Luminescence Spectroscopy, ASTM STP 863*, L. J. Cline Love and D. Eastwood, Eds., American Society for Testing and Materials, Philadelphia, 1985, pp. 16–25.

ABSTRACT: The photo-excitation steps in 3-hydroxyflavone leading to proton-transfer are described. The consequence of the ultra-rapid (<8 ps) phototautomerization is the emission of a unique yellow-green tautomer fluorescence (λ_{max} 525 nm) in hydrocarbon solution, instead of the violet fluorescence expected from the first UV absorption (onset λ 370 nm, first peak λ 354 nm). The tautomerization is intramolecular, involving the transfer of the hydroxyl hydrogen to the neighboring carbonyl group of the 3-hydroxyflavone molecule.

The analytical applications suggested by the qualitative spectroscopic features investigated result from the extreme sensitivity of the intramolecular H-bond to external solvent perturbations, which appear most strikingly in low-temperature spectroscopic studies as induced violet fluorescences of the H-bonded solvates. Using solute concentrations of 10^{-5} to 10^{-7} M, H-bonding impurities are detectable at substoichiometric concentrations. In the case of water in hydrocarbon solvents, the present experiments suggest that 10^{-7} to 10^{-9} M water may be detectable.

Ethers, alcohols, water, and other H-bonding solvents give characteristic violet solvate fluorescence band contours and peak positions, permitting qualitative discrimination of trace contaminants. The use of 3-hydroxyflavone as a fluorescence probe for solvent impurities is suggested. An application is made of isolated-site crystal matrix spectroscopy to determine the intrinsic low-temperature spectroscopic behavior of solutes, such as quinones and ketones, which can H-bond with trace impurities.

KEY WORDS: photon transfer, hydrogen bonds, spectroscopy, fluorescence, hydrocarbons, Shpolskii matrix.

[1]Department of Chemistry and Institute of Molecular Biophysics, Florida State Univeristy, Tallahassee, FL 32306.

Tautomerization of molecular species in the ground state is a very well-known phenomenon. Excited-state tautomerization has recently come under intensive investigation by spectroscopists.

Intramolecular excited state proton transfer is especially interesting because of the wealth of new observations afforded by the integration of techniques of steady state molecular spectroscopy with pulsed laser techniques. New avenues of research and application are opening up in analytical spectroscopy using these new phenomena.

It is clear that for proton translocation in a molecule, various molecular mechanisms may be involved depending on the distance of translocation. In the case of 3-hydroxyflavone [1–4], the intramolecular proton transfer involves a very small motion of the $-OH$ proton. Figure 1 shows the schematic molecular structures of the normal and tautomer molecule and the potential energy diagram for the excitation $S_1 \leftarrow S_0$, proton transfer $S_1 \rightarrow S_1'$, tautomer fluorescence $S_1' \rightarrow S_0'$, and relaxation $S_0' \rightarrow S_0$ for 3-hydroxyflavone (primed states refer to those of the tautomer species).

The dramatic feature of this particular proton-transfer system is the large energy shift, $\sim 10\,000$ cm^{-1}, between the normal molecule absorption $S_1 \leftarrow S_0$ and the tautomer fluorescence $S_1' \rightarrow S_0'$. Figure 2 shows the ultraviolet (UV) absorption spectrum of 3-hydroxyflavone in methylcyclohexane solution at 293 K and the uniquely observed yellow-green fluorescence emission of the tautomer. The large spectral shift is attributed to the extreme electron re-

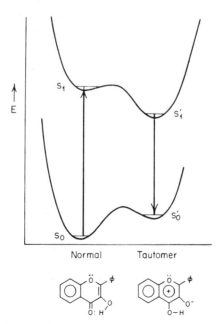

FIG. 1—*Schematic potential energy diagram for proton-transfer spectroscopy of 3-hydroxyflavone.*

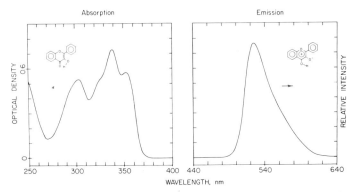

FIG. 2—*Room temperature* (left) *absorption and* (right) *fluorescence spectra of 3-hydroxyflavone in methylcyclohexane solvent (293 K) showing absorption of the "normal" species and unique fluorescence emission of the proton-transferred tautomer species.*

arrangement of the molecular structure upon proton transfer. The carbonyl group becomes more basic in the S_1 excited state, and the aryl-OH group becomes more acidic, and if the intramolecular H-bond is present, rapid proton transfer occurs.

The tautomer is expected to have a benzo-pyrillium electronic structure [1], in analogy to the anthocyanine dyes formed by protonation of a carbonyl leucobase. In the intramolecular excited-state proton-transfer case, the protonation of the carbonyl is light induced (Fig. 1). Picosecond (ps) laser studies of the proton transfer $S_1 \rightarrow S_1'$ in 3-hydroxyflavone indicate that the excited-state tautomerization occurs in <8 ps [5]. Thus, triplet state population and other radiationless processes from the S_1 state are severely competed against, and their role is effectively reduced to zero by the proton-transfer process. Thus, the efficient formation of excited tautomer offers a new realm of tautomer spectroscopic study: fluorescence rise and decay of the tautomer species, kinetics of relaxation to the normal molecule ground state via $S_1' \leftarrow S_0'$ transient absorption studies, and pulsed laser higher state generation, $S_2' \leftarrow S_1'$ [6,7]. Recently the development of a new four-level laser based on 3-hydroxyflavone proton-transfer fluorescence has been described [8,9].

Sensitivity to Specific H-Bonding Interactions

The analytical application which comes to the forefront in the proton-transfer spectroscopy of 3-hydroxyflavone results from the extreme sensitivity in this case of the phototautomerization to H-bonding impurities in the solvent. It has been demonstrated recently [5,10–12] that the proton-transfer tautomerization upon excitation of 3-hydroxyflavone in dilute solution in extremely dry hydrocarbon solvents is independent of temperature from the liquid solution at 298 K to the rigid glass solution at 77 K. Previous publications [1–4] had indicated that at 77 K the tautomerization was inhibited or stopped, with the appearance of the normal molecule violet fluorescence. Thus, the yellow-green tautomer fluo-

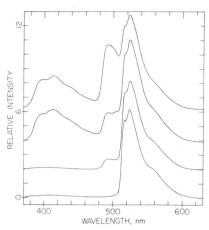

FIG. 3—*Fluorescence spectra of* 2.0×10^{-5} *M 3-hydroxyflavone in methylcyclohexane glass at 77 K as of function of addition of traces of water. The lower curve is for the highly purified anhydrous solution; the highest curve is for solution saturated by the addition of a drop of water. Interference with tautomer fluorescence by H-bonding solvates is evident.*

rescence is the intrinsic property of the 3-hydroxyflavone, and the violet fluorescence arises from solvates, which may interfere with the intramolecular proton transfer process.

Figure 3 illustrates the effect of adding traces of water to the extremely dry hydrocarbon solvent containing 2.0×10^{-5} M solute. The fluorescence curves shown are representative of a whole progressive series [10,11] from the behavior of 3-hydroxyflavone in a rigorously anhydrous hydrocarbon solvent to water-saturated solvent. Brief exposure of an anhydrous solution to the atmosphere at 50% humidity is sufficient to show a perturbation effect of the trace of added water. All of the curves of Fig. 3 are for 77 K hydrocarbon glass, which limits the ultimate solubility of water, but enhances the solvation of 3-hydroxyflavone by the temperature shift of the H-bonding equilibrium.

One analytical application of proton-transfer spectroscopy lies in the possible detection of substoichiometric traces of H-bonding impurities in hydrocarbon solvents. The extreme sensitivity of this method is exemplified by the fact that previous researchers used spectro-grade solvents assumed to be very dry and free of alcohols, ethers, and other H-bonding impurities; yet these very investigations [1-4] were faulted in part by the dominant effect of the impurities at low temperatures (<200 K). Whereas such trace impurities were not phenomenologically manifested at 293 K, at lower temperatures the solvation equilibrium had shifted and the solvate fluorescence properties became dominant.

Figure 3 shows three distinct regions of fluorescence, each assignable to a different molecular species simultaneously present in the low-temperature solution. The unique yellow-green tautomer fluorescence of the internally H-bonded molecule is gradually competed for by the appearance of two new regions of fluorescence, the violet fluorescence of the normal molecule solvate, and a 490-nm band attributed to anion formation in the solute by excited-state proton transfer to water clusters at high water concentration [10,11]. This latter band is not observed in solutions that contain alcohols or ethers as the perturbing impurities [11].

First, it is notable that the 10^{-5} M solute solution shows indications of substoichiometric water concentrations, 10^{-7} M water appearing to be detectable. When a 10^{-7} M solution of 3-hydroxyflavone is used, the detection of 10^{-9} M water would appear to be possible. Our researches were only semi-quantitative since we were interested primarily in outlining the qualitative features of the solvent perturbation behavior. A careful quantitative study should make this a method of ultimate sensitivity for detection of ultra-micro traces of water, alcohols, and ethers in hydrocarbon solvents.

Second, it is evident that a whole series of hydrates can be detected, each having a distinctive perturbation influence on the proton-transfer phenomenon [10,11]. The excitation spectra for the proton-transfer phenomenon (Fig. 4) monitored in the three different fluorescence regions of Fig. 3, indicate that three different equilibrium ground-state molecular species exist, each giving rise to a characteristic emission region: the nonsolvated species yielding the 523-nm tautomer fluorescence band; a solvated species yielding the 495-nm anion fluorescence band; and the chain polysolvated species yielding the 430-nm violet fluorescence band of the "normal" molecule.

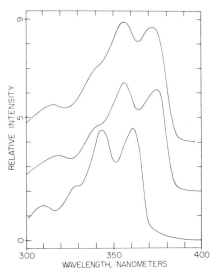

FIG. 4—*The 77 K excitation spectra of 3-hydroxyflavone in methylcyclohexane under various conditions:* (lower) *excitation spectrum of tautomer fluorescence monitored at 523 nm, 3.0×10^{-6} M;* (middle) *excitation spectrum monitored at 495 nm in water-saturated solution, 5×10^{-5} M; and* (upper) *excitation spectrum of "normal" emission in commercial spectroscopic quality solvent, monitored at 430 nm, $\sim 10^{-5}$ M.*

The 3-hydroxyflavone in very dry hydrocarbon is internally H-bonded. Yet the five-membered —C—O—H:O=C— H-bonded ring offers weak H-bonding and thus is highly susceptible to external perturbations. If separate water chains are H-bonded to the carbonyl oxygen on one hand and the hydroxy oxygen or hydrogen on the other, the proton transfer process may be inhibited with the appearance of the normal molecule solvate violet fluorescence.

Qualitative Discrimination of Trace H-Bonding Impurities

The second feature of interest in the application of proton-transfer fluorescence to analytical spectroscopy is the qualitative discrimination of specific impurities which is evident from the spectroscopic behavior of 3-hydroxyflavone in different solvents, and with different trace impurities in hydrocarbon solvents.

The fluorescence behavior of 3-hydroxyflavone in methyl alcohol and in ethyl ether (both at 77 K) are shown in Fig. 5. The spectra show distinctly different fluorescence band contours of the normal molecule solvate violet fluorescence and recognizably distinct fluorescence band maxima. A minor component of the green tautomer fluorescence still is seen in both cases. At 298 K the ratio of the tautomer fluorescence to normal molecule solvate fluorescence is significantly increased in both solutions [1], as thermal activation over the solvation barrier is increased [10,11].

The presence of trace concentrations of methanol or ethyl ether in hydrocarbon solutions (for example, methylcyclohexane, 2-methylbutane) of 3-hydroxyflavone produces distinctive band contours and peak positions analo-

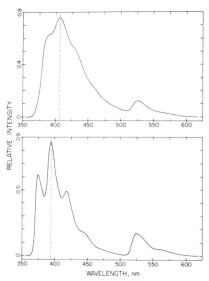

FIG. 5—*Fluorescence emission spectra of 3-hydroxyflavone in* (upper) *methanol and* (lower) *ethyl ether at 77 K. Left: "normal" molecule fluorescence (violet);* right: *tautomer fluorescence (green).*

gous to those observed in the pure solvents (Fig. 5). Thus, a simple qualitative distinction between impurity species present in solution may be made. In fact, for the series ethyl ether, ethanol, methanol, water, a roughly linear correlation is observed between the emission maximum of the violet band (in cm^{-1}) and the H-bonding ability [13] of the impurity molecule.

At 77 K the best spectro-grade commercial hydrocarbon solvents show characteristic violet fluorescence peaks arising from the normal molecule solvate; at 293 K the perturbation effect is not observed, as the solvation equilibrium favors the dissociated species at this higher temperature [5,10,11]. The character of the violet fluorescence band contour also permits an appraisal of whether a single trace contaminant is present (water, alcohol, or ether), or whether several contaminants are present simultaneously.

Using 3-Hydroxyflavone as a Fluorescence Probe

Our research has demonstrated the high sensitivity of the proton-transfer fluorescence of 3-hydroxyflavone to hydrocarbon solvent impurities when used as low-temperature glass solvents. The growth of interest in low-temperature spectroscopy as an analytical tool, especially for phosphorescence (triplet state) studies of organic molecules, focuses attention on the behavior of such systems.

The isolated-site crystal matrix (Shpolskii matrix) is a low-temperature technique that warrants comparison at this point. The Shpolskii technique consists of freezing dilute solutions of a given solute in a series of n-alkane solvents of increasing chain length. At a certain chain length, the solute may replace one or more solvent molecules in an isomorphic substitution in the crystal lattice site. At that particular solute/solvent combination, the spectrum dramatically sharpens (even at 77 K; with line-like sharpening at 20 and 4 K) owing to the diminution of inhomogeneous spectral broadening, even though the solid solution is polycrystalline. For 3-hydroxyflavone a typical result is shown in Fig. 6 for n-heptane solvent [12].

The importance of the isolated-site crystal matrix (Shpolskii) spectrum for the present discussion is that it isolates the solute from solvent impurities since the crystal formation can set in before the solvation equilibrium becomes significant. Thus, the intrinsic behavior of a solute molecule is often revealed by a study of the Shpolskii matrix spectrum.

Standard criteria for spectroscopic purity focus primarily on the presence of absorbing and emitting impurities, very often with little concern for such species as water, alcohols, and ethers, each of which is common to most laboratory environments. The dramatic spectral changes that occur in the 3-hydroxyflavone system with solvent purification establish the consequences of their presence in that system. The elimination of these contaminants is, however, no simple task [11]. For example, following the introduction of ethyl ether to a grease-free vacuum system, an ether impurity was observed in methylcyclohexane solutions of 3-hydroxyflavone for several weeks before the source of the contaminant was

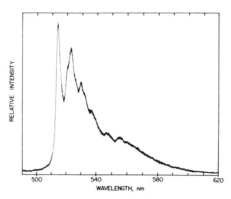

FIG. 6—*The 77 K tautomer fluorescence spectrum of 3-hydroxyflavone in the polycrystalline n-heptane Shpolśkii matrix (as an isolated site crystal matrix). This spectrum indicates the intrinsic spectroscopic behavior of 3-hydroxyflavone in the absence of H-bonding interference.*

discovered. The ethyl ether was apparently leaching out of stop cock O rings. It is not unexpected then that several other low-temperature spectroscopic studies may have unwittingly been interfered with by the presence of such H-bonding impurities. In particular, the observation of nonexponential phosphorescence decay for many aromatic ketones, quinones, and azines has remained a source of confusion in the literature for several years [14–19].

The use of 3-hydroxyflavone as a fluorescence probe for trace solvent impurities can be carried out in parallel to the spectroscopic study of another solute in the same solvent. The 3-hydroxyflavone "reference" provides information as to the nature and the concentration of any impurity species that is present in the solvent, or is otherwise introduced in the preparation procedure. The correlation of this data with experimental observables of the system under investigation, such as the degree of nonexponentiality of the decay curve, for example, may be utilized to determine whether the apparently anomalous behavior originates from interactions with solvent impurity molecules.

In this laboratory we have successfully used this technique to elucidate the origin of the nonexponential phosphorescence decay of xanthone in 77 K hydrocarbon glasses [20]. The recent very elegant work by Griessler and Bramley [21] and Connors and Christian [22] has successfully utilized the Shpolśkii technique to unravel the (rather complicated) intrinsic properties of the xanthone molecule. The apparently anomalous behavior of xanthone in 3-methylpentane glass however remained unresolved [18,19]. Xanthone is particularly susceptible to H-bonding perturbations because of the near degeneracy of the two lowest triplet states, T_1 (π, π^*) and T_2 (n, π^*). This energy gap is only 27 cm^{-1} in a *n*-hexane crystal matrix [21]. Our studies [20] clearly indicate that the slower component of the xanthone phosphorescence decay originates from xanthone molecules that are associated with solvent impurities.

Conclusion

This survey of the spectroscopic behavior of 3-hydroxyflavone illustrates the diversity of spectral phenomena that may be observed when intramolecular proton transfer can occur. In the case of 3-hydroxyflavone the H-bond between the −OH hydrogen and the neighboring carbonyl group permits the proton transfer. The presence of the γ-pyrone ring allows for aromatization in the tautomer, with a 10 000-cm^{-1} spectral shift of the tautomer fluorescence relative to the first singlet-singlet absorption.

The analytical applications of this system depend on the wide range of spectroscopic effects, which are observed with solvation by H-bonding solvents, or as trace impurities. Water, alcohols, and ethers all perturb the intramolecular H-bond, resulting in an inhibition of the tautomerization process and the subsequent observation of the normal molecule solvate violet fluorescence. This effect appears most strikingly in low-temperature glassy solutions in which the ground state equilibrium favors the formation of H-bonded complexes. Also, in these high-viscosity systems the various solvent reorganizational processes that allow for relaxation from the excited-state normal molecule solvate to the tautomer species are eliminated.

The observation of distinct tautomer and "normal" molecule fluorescences in different spectral regions permits easy discrimination between the various ground state species. Thus, the ultimate sensitivity of this system for the detection of trace H-bonding impurities is limited only by the limits of detection of the 3-hydroxyflavone fluorescence emission.

Other molecular systems with a similar electronic mesomerization may be investigated. The necessary conditions for excited-state proton transfer in this class of molecules are (1) an oxo-ring with aliphatic conjugation, (2) an H-bonded hydroxy-to-carbonyl, and (3) favorable state energies for excitation. It is of interest to point out that in the present case of 3-hydroxyflavone, the components (1) and (2) are on the same ring, as is also the case in the polyhydroxylated flavone quercetin and other related flavones. However, in the leuco-base to anthocyanine tautomerization, as a phototautomerization, the H-bonded OH $\cdots$ O=C combination is in the side phenyl ring (B ring) while the oxo-group is in the A ring. Various electronic systems can thus be investigated to parallel the photo-induced pyrillium tautomer discussed here.

References

[1] Sengupta, P. A. and Kasha, M., *Chemical Physics Letters,* Vol. 68, 1979, pp. 382–385.
[2] Woolfe, G. W. and Thistlethwaite, P. J., *Journal of the American Chemical Society,* Vol. 103, 1981, pp. 6919–6923.
[3] Itoh, M., Tokumura, K., Tanimoto, Y., Okada, Y., Takeuchi, H., Obi, K., and Tanaka, I., *Journal of the American Chemical Society,* Vol. 104, 1982, pp. 4146–4150.
[4] Strandjord, A. J. G., Courtney, S. H., Friedrich, D. M., and Barbara, P. F., *Journal of Physical Chemistry,* Vol. 87, 1983, pp. 1125–1133.
[5] McMorrow, D., Dzugan, T., and Aartsma, T. J., *Chemical Physics Letters,* Vol. 103, 1984, pp. 492–496.

[6] Itoh, M. and Fujiwara, Y., *Journal of Physical Chemistry,* Vol. 87, 1983, pp. 4558–4560.
[7] Itoh, M., Tanimoto, Y., and Tokumura, K., *Journal of the American Chemical Society,* Vol. 105, 1983, pp. 3339–3340.
[8] Khan, A. U. and Kasha, M., *Proceedings of the National Academy of Science,* Vol. 80, 1983, pp. 1767–1770.
[9] Chou, P. T., McMorrow, D., Aartsma, T. J., and Kasha, M., *Journal of Physical Chemistry,* Vol. 88, in press.
[10] McMorrow, D. and Kasha, M., *Journal of the American Chemical Society,* Vol. 105, 1983, pp. 5133–5134.
[11] McMorrow, D. and Kasha, M., *Journal of Physical Chemistry,* Vol. 88, 1984, pp. 2235–2243.
[12] McMorrow, D. and Kasha, M., *Proceedings of the National Academy of Science,* Vol. 81, 1984, pp. 3375–3378.
[13] Cramer, L. E. and Spears, K. G., *Journal of the American Chemical Society,* Vol. 100, 1978, pp. 221–227.
[14] Griffin, R. N., *Photochemistry and Photobiology,* Vol. 7, 1968, pp. 175–187.
[15] Yang, N. C. and Murov, S., *Journal of Chemical Physics,* Vol. 45, 1966, p. 4358.
[16] Wagner, P. J., May, M. J., Haug, A., and Graber, D. R., *Journal of the American Chemical Society,* Vol. 92, 1970, pp. 5269–5270.
[17] Kanda, Y., Stanislaus, J., and Lim, E. C., *Journal of the American Chemical Society,* Vol. 91, 1969, pp. 5085–5089.
[18] Pownall, H. J. and Huber, J. R., *Journal of the American Chemical Society,* Vol. 93, 1971, pp. 6429–6436.
[19] Pownall, H. J., Connors, R. E., and Huber, J. R., *Chemical Physics Letters,* Vol. 22, 1973, pp. 403–405.
[20] McMorrow, D. and Irons, M., to be published.
[21] Griesser, H. J. and Bramley, R., *Chemical Physics,* Vol. 67, 1982, pp. 361–371 and pp. 373–389.
[22] Connors, R. E. and Christian, W. R., *Journal of Physical Chemistry,* Vol. 86, 1982, pp. 1524–1528.

Raymond F. Chen[1] *and Carrie H. Scott*[1]

Global and Nonglobal Rotations in Proteins Detected by Fluorescence Polarization

---

REFERENCE: Chen, R. F. and Scott, C. H., "**Global and Nonglobal Rotations in Proteins Detected by Fluorescence Polarization,**" *Advances in Luminescence Spectroscopy, ASTM STP 863*, L. J. Cline Love and D. Eastwood, Eds., American Society for Testing and Materials, Philadelphia, 1985, pp. 26–39.

ABSTRACT: Fluorescence polarization methods have been used to assess protein motion for about three decades. These methods are still evolving, but basically can be divided into three classes: (1) steady state methods, (2) time-resolved anisotropy measurement, and (3) differential polarized phase fluorometry. A brief overview of these methods is given, and data are presented on proteins that are unlabeled, or conjugated with dansyl (1-dimethylaminonaphthalene-5-sulfonyl) group. The experiments show that the dansyl label exhibits a fast thermally activated motion, which is viscosity independent, as well as a slower motion characteristic of the global rotation of the protein as a whole. Data on the intrinsic ultraviolet fluorescence of proteins show that the global rotation rate can be obtained from steady state data, which also can detect independent motion of the tryptophan residues. Time-resolved anisotropy data are presented showing directly the rapid rotation of the dansyl group. These data illustrate the ease with which polarization data detect global and nonglobal rotations in proteins.

KEY WORDS: proteins, fluorescence, molecular relaxation, protein fluorescence, fluorescence polarization, fluorescence anisotropy, protein flexibility, Perrin plot, dansyl fluorescence, tryptophan fluorescence, fluorescence lifetime

---

A detailed picture of protein structure was afforded by the magnificent contributions of X-ray crystallography, which offered the hope that the functional mechanisms of proteins could be known from the positions of the individual atoms. The beauty of the X-ray model of structure obscured the fact that it is basically static, deriving as it does from the rigidity and order of crystals. Many observations cannot be explained on the basis of an unchanging, unique protein conformation. In recent years, it has become apparent that rapid fluctuations of

---

[1]Medical officer and biologist, respectively, National Heart, Lung, and Blood Institute, Building 10, Room 5D-18, Bethesda, MD 20205.

parts of proteins, on the time scale of nanoseconds and faster, are important properties. The idea that parts of proteins are flexible and in constant motion helps to explain the action of antibodies and enzymes, how proteins undergo transitions between conformational states, the formation of fibrin clots, and many binding phenomena. Only the fact that proteins "breathe" can explain hydrogen exchange with residues in interior regions shown to be "anhydrous" by X-ray crystallography [1,2]. Penetration of oxygen (O_2) to "buried" tryptophans can be shown by fluorescence and phosphorescence quenching [3,4]. Fluctuations of proteins can also be demonstrated by electron spin resonance (ESR) [5] and nuclear magnetic resonance (NMR) methods [6,7]. However, the method, which in retrospect may have afforded some of the first demonstrations of the dynamic fluctuations occurring in proteins, is fluorescence polarization.

Fluorescence Polarization of Proteins — Historical Note

Introduced into protein chemistry in 1952 by Weber [8], the fluorescence polarization method almost immediately showed that proteins were flexible. In a 1967 review [9], tabulated results of fluorescence polarization studies on various proteins labeled with dyes indicated that more than half of the relaxation times reported were faster than expected for overall rotational diffusion. While some workers were inclined to attribute these findings to rotation of the probe about the bond by which it was attached to the protein, a trivial feature, this has not been confirmed. On the other hand the potential for rotation of a segment that includes the attachment bond is likely, since many labels attach at ε-amino groups of lysyl residues, creating the possibility of a freely rotating, long hydrocarbon chain as shown in Fig. 1. This type of segmental rotation could be thought of as an artifact of the use of extrinsically added labels. However, it could be argued that since salt bridges are equilibrium structures, flexible and rotating lysyl sidechains occur at least some of the time in the native structure. From this point of view one would have to say that the dye is accurately reporting a naturally occurring subunit rotation.

The fluorescence polarization method has advanced considerably in theory and technique in recent years. The original method used steady state radiation of the sample and is still extremely useful. Instrumental developments have permitted polarization to be observed as a function of time (on the nanosecond scale) following pulsed excitation. Still another method uses sinusoidally modulated (at megaHertz frequencies) polarized excitation, and rotational information is extracted from the phases of the polarized components of the emission. Also much theoretical work has been performed to calculate the depolarization characteristics of isotropic and anisotropic rotators undergoing free and hindered motion. The present communication offers a brief review of the method and gives examples showing how depolarization data distinguishes between rotation of the entire macromolecule ("global rotation") and motion of subunits ("nonglobal rotation").

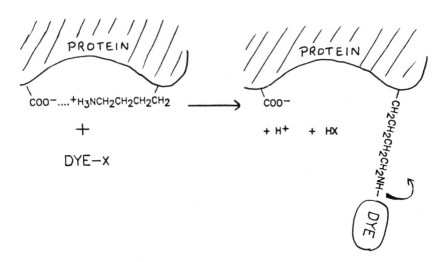

FIG. 1—*Labeling of a protein by a dye. Attachment is visualized as occurring at the ε-amino group of a lysyl residue, converting the sidechain into a freely rotating group.*

Background and Theory

Static (Steady State) Polarization

If a sample is continuously irradiated with linearly polarized light and the emission is observed at right angles to the direction of the exciting beam

$$P = (I_v - I_h)/(I_v + I_h), \quad A = (I_v - I_h)/(I_v + 2I_h), \quad \text{and}$$
$$P = 3A/(2 + A)$$

where P and A are the fluorescence polarization and anisotropy, respectively, and I is the intensity observed though analyzing polarizers oriented vertically v or horizontally h, corresponding to directions parallel and perpendicular to the direction of excitation polarization.

Perrin [10] derived the relation between the limiting (maximum) polarization P_o, the polarization P, and the fluorescence lifetime

$$1/P - 1/3 = (1/P_o - 1/3)(1 + 3\tau/\rho_o) \tag{1}$$

where ρ_o is the rotational relaxation time of a sphere and τ is the fluorescence lifetime. The latter quantity could also be calculated from the Einstein-Stokes equation

$$\rho_o = 3\eta V/kT \tag{2}$$

where η is the viscosity, V the molecular volume, k is the molecular gas constant, and T is the absolute temperature. From the above, it follows that

$$1/P - 1/3 = (1/P_o - 1/3)(1 + RT\tau/\eta V) \tag{3}$$

Weber [8,11] applied Perrin's theory to the case of proteins labeled with fluorescent dyes. By assuming that proteins behaved as ellipsoids and that labeling was random and rigid, a modified Perrin equation permitted determination of protein relaxation rates

$$1/P - 1/3 = (1/P_o - 1/3)(1 + 3\tau/\rho_h) \quad (4)$$

where ρ_h is the mean harmonic rotational relaxation time of an ellipsoid

$$\rho_h = 1/\sum (1/\rho_i) \quad (5)$$

where ρ_i is the rotational relaxation time associated with axis i.

The Perrin-Weber equation, in terms of the anisotropy A and the limiting anisotropy A_o has the simple form

$$A_o/A = 1 + 3\tau/\rho_h \quad (6)$$

Experimentally, in order to determine the relaxation time, one usually obtains polarization measurements at different temperatures or viscosities. A Perrin plot, consisting of $1/P$ versus $T\tau/\eta$, will normally give a linear extrapolation to $1/P_o$ at $T\tau/\eta = 0$. From this, ρ_h is calculated using Eq 4.

Weber [12] calculated ρ_h for ellipsoids having different axial ratios. This raised the possibility that shape information could be obtained. There were, however, difficulties with the method.

Equation 4 assumes a strictly random orientation of dyes. If only one or a few labels are attached, the fluorescence may reflect preferential orientation of the absorbing and emitting dipoles vis-a-vis the ellipsoidal axes. For a single label attached to ellipsoids of revolution, Teale and Badley [13] have derived an equation for calculation of ρ_h/ρ_o as a function of axial ratio and the angle between the emission dipole and the axis of symmetry (assuming coincidence of the absorption and emission dipole directions). A common error is for rotational relaxation time measurements to be made with protein conjugates having a very low degree of labeling.

The Perrin plot ceases to be linear in the case of multiple relaxation times. Perhaps the earliest treatment of the effect of segmental rotations on fluorescence depolarization data was that of Gottlieb and Wahl [14] who applied their theoretical results to data from synthetic polymers. Intuitively, one can see that if the global relaxation is slow compared to a segmental rotation, the Perrin plot will be concave towards the x axis. If the segmental rotation is both rapid and unhindered, essentially all polarization will be lost at higher values of T/η even if the global relaxation is slow. With a rapid but hindered rotation, the effect of the segmental rotation will be to cause a partial depolarization. The effect will be to reduce P_o, but the global relaxation rate can still be obtained. Figure 2 illustrates these concepts.

An apparent limiting polarization P_o is obtained by extrapolation of the Perrin plot, and from this one calculate the angle over which the flexible subunit can

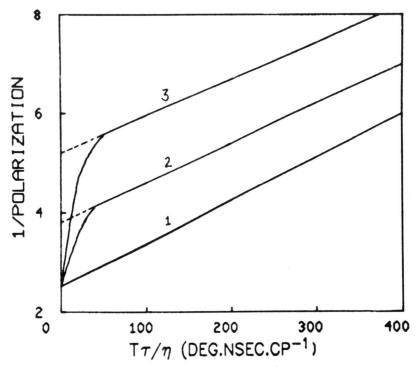

FIG. 2—*Idealized Perrin plots. Curve 1 supposes a randomly labeled protein behaving as a sphere and exhibiting only global relaxation. Curve 2 assumes a subunit rotation that is rapid and hindered. Curve 3 assumes a less hindered subunit rotation than in Curve 2.*

move. More conveniently expressed in terms of anisotropy, the relation is

$$A'_o/A = (3\cos^2\theta - 1)/2 \qquad (7)$$

where A'_o is the apparent limiting anisotropy and θ is the average angle of the cone describing the range of motion of the subunit [14].

For two or more rotational rates, which are closer together, the curved Perrin plots cannot unambiguously yield relaxation rates [15,16]. Brochon and Wahl [16] have calculated Perrin plots for different combinations of rotational rates in a study of dansylated gamma globulin. If two relaxations are considered with weighting factors w_1 and w_2, the relation of the two rates ρ_1 and ρ_2 is most easily expressed in terms of anisotropies

$$A = A_0[w_1/(1 + \rho_1)] + [w_2/(1 + \rho_2)] \qquad (8)$$

Even this equation does not hold if the rotational rates are close together [17].

Time-Resolved Fluorescence Anisotropy

Modern instrumentation permits observation of the rate of decay of fluorescence polarization as a function of time after a short pulse of polarized excitation. The literature on time-resolved measurements of this type utilize A instead of P, as in the steady state method. Also the rotational correlation time ϕ is normally used instead of the relaxation time ρ, the relation between them being $\rho = 3\phi$.

In this method, the sample is ideally excited by a light pulse having a negligible width, and the vertically and horizontally polarized components of the emission are followed as a function of time. $A(t)$ calculated from such data will in the case of a sphere be related to ϕ as follows

$$A(t) = A_o e^{-t/\phi} \tag{9}$$

Thus, a plot of ln A versus time yields ϕ directly.

The time resolved anisotropy method has the advantage over the static method that the relaxation rate is determined without altering T or η. Note that in the case of an ideal sphere, Eq 9 is independent of the lifetime τ.

In the general case of an irregular body rather than a sphere, with a single emission oscillator, the expression for $A(t)$ is highly complex. For rotation in isotropic media, the rigorous expression consists of five exponentials [18–21]. For ellipsoids of revolution, $A(t)$ is given by three exponentials, and one exponential is required only in the case of a sphere. The calculation of expressions for $A(t)$ for shapes containing hinges and various degrees of flexibility has become somewhat of a mathematical challenge to which both analytical [22] and computer methods [23,24] have been applied.

In the case of fluorophors embedded in anisotropic milieus, such as lipid membranes, rotational freedom may be limited. In this case, A decays to a value significantly greater than zero at times that are long relative to the relaxation time. Expressions for $A(t)$ have been calculated in terms of the angles between the absorption and emission dipoles and the angles they make with the symmetry axis of an ellipsoid, the degree of rotational freedom, the relaxation rate, and other parameters [25–28].

When the observed semilog plot of the emission anisotropy versus time is nonlinear, there is no a priori reason to assume the presence of more than one rotating unit, since a monoexponential anisotropy decay is expected only for spheres. Nonetheless, practical experience has shown that most homogeneous protein systems have anisotropy decay plots approximating that for a sphere, so the harmonic mean rotational correlation time ϕ_h, analogous to ρ_h, is calculated from a monoexponential expression such as Eq 9. Nonexponential curves for ln A versus t were observed by Yguerabide et al [29] for immunoglobulin and were resolved into different spherical rotators by assuming that the curve was a sum of two anisotropy decays

$$A(t) = A_o(a_1 e^{-t/\phi_1} + a_2 e^{-t/\phi_2}) \tag{10}$$

where a and ϕ are the weighting factors and rotational correlation times. While such a procedure lacks a rigorous mathematical justification, it seems to yield qualitatively reasonable descriptions of segmental rotations [29,30].

Differential Polarized Phase Fluorometry

In the phase shift method of determining the fluorescence lifetime, a sample is irradiated with a sinusoidally modulated beam. The fluorescence is delayed relative to the excitation, and the lifetime can be calculated from the amount of delay, expressed as a phase shift. If polarized excitation is used and the vertical and horizontal components of the emission are monitored, they will have different phase shifts if the solution has polarized emission. Since the data contain the same information as that obtained by pulse methods, it is not surprising that the rotational rate can be calculated. The general equation can be written as [31]

$$\tan \Delta = [(2R\tau)\omega\tau A_o]/[m(1 + \omega^2\tau^2) + (2R\tau/3)(2 + A_o) + (2R\tau)^2] \quad (11)$$

where $m = (1 + 2A_o)(1 - A_o)/9$, Δ = the difference in phase angle of the two polarized emission components, R = rotational rate in rad/s, ω = circular modulation frequency, and A_o and τ are the limiting anisotropy and lifetime.

The detailed theory of this method has been given by Weber [32]. At the present time, the method has been applied to detecting and quantifying anisotropic and hindered rotations [$33,34$]. Since the method is quite new, relatively little literature using it has become available, but more data can be expected from this interesting technique.

Observations of Segmental Rotations in Proteins

Since polarization studies on proteins are often done with the extrinsic fluorescence of dye labels, it was of interest to know if it was a general rule that the dye had degrees of rotational freedom greater than possessed by the protein as a whole. In other words, if the dye label rotated as indicated in Fig. 1, dye conjugates themselves would serve as models of segmental flexibility. The form of Perrin plots for such conjugates should differ sharply from those where the fluorophor is firmly attached, such as in the case of the intrinsic tryptophans or dyes attached in some crevice.

Static fluorescence polarization measurements were performed on dansylated proteins (labeled by reaction with 1-dimethylaminonaphthalene-5-sulfonyl chloride) and unlabeled proteins. The results given below suggest that thermally activated rotations of the dansylated segment occur as a rule in contrast to the native protein fluorescence, which reflects global protein rotation.

Materials and Methods

All proteins were crystalline preparations obtained from commercial sources. Serum albumins were charcoal treated to remove impurities [35]. Dansyl chloride was purchased from Pierce Chemical Co. Dansyl proteins were prepared by reaction of dansyl chloride dissolved in acetone added to proteins in $0.1M$ sodium bicarbonate ($NaHCO_3$). Typical reaction times were 1 to 2 h, after which the protein was separated from free dye by passage through Sephadex G-25. Polarization measurements were performed on solutions containing 0.1% protein in a 0.03M potassium phosphate buffer, pH 7.4.

Static polarization measurements were made on an Aminco-Bowman spectrofluorometer fitted with Polacoat and Polaroid® polarizers [36]. Time resolved intensity and anisotropy data were obtained with an Ortec 9200 system fitted with a Tracor Northern TN-1710 multichannel analyzer and a Photochemical Research Associates, Ltd., (PRA) flashlamp. Fluorescence lifetimes were obtained with a TRW Instruments decay time apparatus [37], which we find gives approximate arithmetic mean lifetimes [30] with multiexponential decays.

Results

Static Polarization Measurements on Dansyl Conjugates — Figure 3 shows data for dansyl conjugates of immunoglobulin G, lysozyme, trypsinogen, and chymotrypsinogen. The Perrin plots for each protein consists of four types of

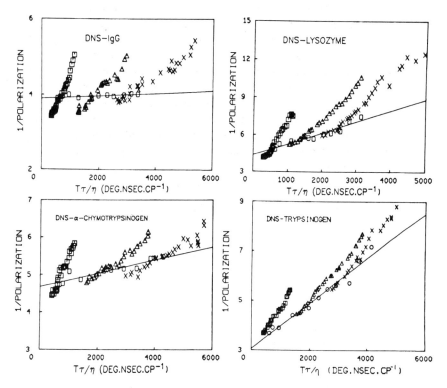

FIG. 3—*Perrin plots for the visible fluorescence of some dansyl labeled proteins. Four sets of data are given for each protein: O data at 25° in solutions of different glycerol concentration; other points are obtained at different temperatures in water X, 20% glycerol △, and 50% glycerol □. The straight lines are drawn through the isothermal points. DNS = dansyl; IgG = immunoglobulin G. The solutions also contained 0.03 M potassium phosphate, pH 7.4.*

data: points obtained isothermally at 25°C by adding glycerol to change the viscosity, and points obtained by heating over the range 8 to 45°C in solutions containing 0, 20, and 50% (by weight) glycerol. Excitation and emission were at 340 and 500 nm. The fluorescence lifetimes used in the Perrin plots were measured with the TRW instrument at 25°C in water and glycerol and assumed to be proportional to the intensity ($I_v + 2I_h$) at any other temperature.

Several features of these Perrin plots are worthy of note: (1) The isothermal data points define a straight line (shown) that intersects the y axis at a point that is well above the actual value corresponding to the P_o for the dansyl group (0.40). (2) If a straight line were drawn through each set of heating data, that line would intersect near the true $1/P_o$, which for the dansyl group is 2.5. (3) In no case do the heating and isothermal curves coincide. This is true not only for the conjugates shown in Fig. 3, but for all dansyl conjugates we have examined. (4) The heating data points sometimes are suggestive of an upward curvature (see dansyl lysozyme in 20% glycerol).

The Perrin plots seem to show that the heating and isothermal data measure the rotations of different species. The fact that the isothermal data extrapolate to a false, or apparent, limiting polarization, which is lower than the true P_o suggests that there is a viscosity independent depolarizing factor. If isothermal lines were drawn for different temperatures, it would be seen that the difference between P_o' and P_o would decrease with temperature and vanish at absolute zero. This depolarizing factor would be a thermally activated local rotation involving the dansyl group. The angle θ through which this rotation moves is easily calculated from Eq 7. Values of θ represent those angles subtended during the excited lifetime; the values of τ and θ for a number of dansyl conjugates are given in Table 1. These data show the average θ values for the dansyl groups attached to a given protein in this series ranged from 20.7 to 35.6°.

TABLE 1—*Rotational freedom of the dansyl groups in various protein conjugates.*[a]

Dansyl Conjugate	Lifetime, ns	θ, degrees
Trypsinogen	12.6	21.6
Bovine serum albumin	17.9	23.1
Lysozyme	9.7	33.4
Chymotrypsinogen	13.0	35.6
Pepsin	12.0	27.8
Immunoglobulin G	10.9	30.8
Human serum albumin	18.0	30.8
Fibrinogen	11.6	33.9
β-Lactoglobulin	20.1	23.9
Ovalbumin	13.0	34.4
Thyroglobulin	13.1	30.6
Liver alcohol dehydrogenase	14.6	20.7

[a] The angle θ over which the dansyl group rotates during the fluorescence lifetime was calculated from Eq. 7, using data from plots such as shown in Fig. 3. The degree of labeling varied from 0.7 to 13 dansyl groups per protein molecule.

As mentioned above, some heating curves have an upward curvature, which would indicate a thermally induced increase in rotational freedom. Although the rotational rate increases with temperature for a rigid structure, the volume should not change. However, using the Perrin equation for a sphere (Eq 4) we can calculate the volume of the equivalent rotating sphere for the polarization data for heating. Results for such calculations are shown in Fig. 4. It can be seen that the volume of the rotating unit decreases with temperature for these examples. One would assume that heating results in a wider range of rotation of the dansylated segment, thus accounting for the apparent smaller rotating volume.

The volume V obtained in this way has little physical meaning other than to show an increasing rotational freedom. The rotating segment is not likely to be either spherical or rotating isotropically.

Static Polarization on Unlabeled Proteins — For each protein examined, Perrin plot data were obtained in the same four ways as for the dansyl conjugates above. Excitation and emission wavelengths were 280 and 350 nm, with bandwidths of 10 nm. Such a wide bandwidth was used for excitation in order to achieve an averaging effect; it is known that the polarization spectrum of tryptophan fluorescence is highly structured and can change under different conditions [38]. The large bandwidth serves to help avoid large changes in P_o caused by small spectral shifts during heating or changes in viscosity.

Examples of Perrin plots for unlabeled tryptophan-containing proteins are shown in Fig. 5. The polarization values are much lower than those for dansyl conjugates, since the P_o for the intrinsic fluorescence is lower, especially when excited at 280 nm. Nevertheless, it can be seen that reasonable extrapolations can

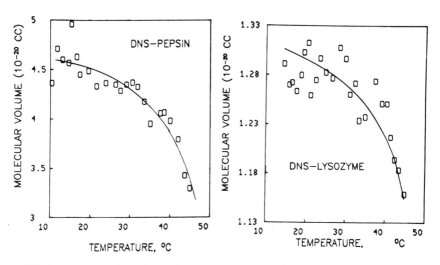

FIG. 4 — *Apparent volume changes for the rotating group in dansyl proteins. Data such as shown in Fig. 3 were used to calculate the molecular volume V from the Perrin equation for a sphere (Eq 3 of text) as a function of temperature. The results shown are for dansyl lysozyme in 20% glycerol (see Fig. 3) and dansyl pepsin in water.*

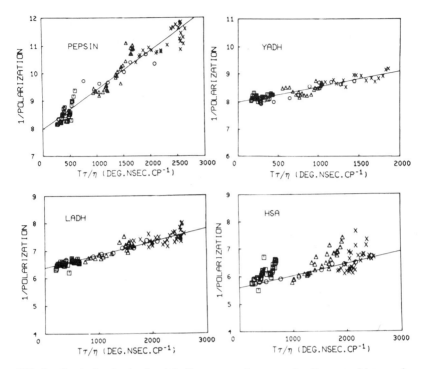

FIG. 5—*Perrin plots for the ultraviolet fluorescence of some proteins. Four sets of data are shown for each protein, with the same symbols as in Fig. 3. The straight lines are drawn through the isothermal data points O. HSA is human serum albumin; YADH and LADH are yeast and liver alcohol dehydrogenases, respectively.*

be made through the isothermal data points to yield on apparent P_o to be used for calculation of relaxation times. The results of such calculations are shown in Table 2.

In comparing Figs. 3 and 5, it is immediately clear that heating the unlabeled proteins does not give as clear evidence of thermally activated rotational flexi-

TABLE 2—*Relaxation parameters from intrinsic protein fluorescence polarization.*[a]

Protein	Mean Lifetime τ, ns	Relaxation Time ρ_h, ns
Human serum albumin	7.5	124
Pepsin	6.4	48.5
Yeast alcohol dehydrogenase	5.3	125.2
Liver alcohol dehydrogenase	5.0	105.1

[a]The relaxation times were calculated from the data of Fig. 5.

bility, as in the case of the dansylated proteins. Tryptophans, being hydrophobic, would not be expected to lie on the surface of most proteins.

We also have noted that excessive scatter is encountered while attempting to make Perrin plots of intrinsic fluorescence by heating alone, even though most of the points obtained fall in the general range of the straight line defined in the isothermal experiment. The reason for this is that there is too short of a span $T\tau/\eta$ accessible by heating alone. The amount of scatter may be caused by a variety of factors, but mainly the fact that one is working in a range of low P values. Volume changes upon heating, as reported by Ross and coworkers [*39*] may also affect the data.

Some suggestions of a faster rotating unit are seen in the heating data for human serum albumin (HSA) (Fig. 4) in spite of the scatter. From pulsed anisotropy data obtained with subnanosecond excitation, Munro et al [*40*] have suggested that HSA's lone tryptophan has a certain degree of independent rotational freedom, in consonance with the present results.

Nanosecond Anisotropy Experiments — It is possible to detect the rapid rotation responsible for the partial depolarization of dansyl fluorescence, which yields low apparent P_o values such as in Fig. 3. Using photon counting time-resolved anisotropy measurements, data such as that shown in Fig. 6 for dansyl ovalbumin conjugate have been obtained. Note that the early part of the anisotropy decay curve shows a rapid drop. The majority of the curve yields, according to Eq 9, a rotational correlation time of 35 ns, reasonable for global rotation. The rapid part of the decrease in ln A represents the hindered motion over an angle θ (Table 1).

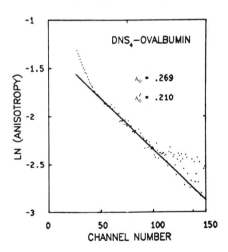

FIG. 6—*Anisotropy as a function of time in a dansyl ovalbumin conjugate. Time base is 1.36 channels per nanosecond. A_o and A_o' are the experimental and apparent (extrapolated) anisotropies at Channel 26, which corresponds to the flash peak.*

Discussion and Conclusions

Fluorescence polarization methods are ideally suited for monitoring the dynamic properties of proteins, since the rotations and lifetimes are both in the nanosecond range. While simple in concept, polarization techniques as commonly practiced use assumptions that must be kept in mind. For instance, both steady state and time-resolved data are usually analyzed with equations that strictly apply to spheres. For multiple labeled proteins, the fluorescence decay may be multiexponential, yet equations are used that assume monoexponential decay. Also, random labeling of proteins is often assumed, even though there is no evidence that this ever occurs. The literature is replete with reports of so-called mean rotational relaxation times for proteins labeled with an average of less than one fluorophor whose angular disposition relative to the symmetry axis is probably fixed and unknown.

The results of depolarization experiments presented here show the following. (1) Dansyl conjugates exhibit both a thermally activated rapid rotation and a viscosity independent rotation, by steady state analysis. (2) The rapid and slow rotations can be visualized directly by the time-resolved anisotropy measurement. (3) Subunit rotation is absent when examining the intrinsic fluorescence of some proteins, but in HSA some rotational freedom of the tryptophan is observable by steady state methods. (4) Global rotations can be obtained from steady state isothermal measurements of intrinsic protein polarization.

The viscosity-independent rotations observed require additional comment. It is clear from the data (Fig. 3) that the isothermal Perrin plots, extrapolated to infinite viscosity, intersect at a value of $1/P_o$ corresponding to a limiting polarization significantly lower than that of the isolated fluorophor in rigid media. That the label can still rotate at high external viscosity simply suggests that the macroscopic viscosity is not what is sensed by the fluorphor. This could result from the label being in a protected pocket insulated from solvent, but this is unlikely since there is spectral evidence that most dansyl groups attach to solvent exposed regions. Alternatively, one could view the protein as an undulating mass, whose global motion is damped by glycerol, but whose internal motions are functions of an effective internal "viscosity." The dansyl groups would be riding the waves of such thermally activated undulations.

Fluorescence polarization methods are still evolving although over 30 years have passed since their introduction into protein chemistry by Weber [8]. The present overview is not to be taken as a review because these techniques have now been applied to so many systems that any comprehensive summary would be impracticably unwieldy. Nevertheless, as technology advances, one expects that even more interesting applications to protein dynamics will appear.

References

[1] Englander, S. W. and Mauel, C., *Journal of Biological Chemistry*, Vol. 247, 1972, pp. 2387–2394.

[2] Englander, S. W. and Rolfe, A., *Journal of Biological Chemistry*, Vol. 248, 1973, pp. 4852–4861.
[3] Lakowicz, J. R. and Weber, G., *Biochemistry*, Vol. 12, 1973, pp. 4171–4179.
[4] Teale, F. W. J. and Badley, R. A., *Biochemical Journal*, Vol. 116, 1970, pp. 341–348.
[5] Johnson, M. E., *Biochemistry*, Vol. 17, 1978, pp. 1223–1228.
[6] Wooten, J. B. and Cohen, J. S., *Biochemistry*, Vol. 18, 1979, pp. 4189–4179.
[7] Highsmith, S., Akasaka, K., Konad, M., Goody, R., Holmes, K., and Wade-Jardetzky, N., *Biochemstry*, Vol. 18, 1979, pp. 4238–4244.
[8] Weber, G., *Biochemical Journal*, Vol. 51, 1952, pp. 145–155.
[9] Chen, R. F. *Fluorescene: Theory, Instrumentation and Practice*, Marcel Dekker, Inc., NY, 1967, pp. 443–509.
[10] Perrin, F., *Journal of Physics of Radium*, Vol. 5, 1934, pp. 497–511.
[11] Weber G., *Biochemical Journal*, Vol. 51, 1952, pp. 155–167.
[12] Weber, G., *Advances in Protein Chemistry*, Vol. 8, 1953, p. 415.
[13] Teale, F. W. J. and Badley, R. A., *Biochemical Journal*, Vol. 116, 1970, pp. 341–348.
[14] Gottlieb, Y. and Wahl, P., *Journal of Chemical Physics*, Vol. 60, 1963, pp. 849–856.
[15] Wahl, P., in *Biochemical Fluorescence: Concepts*, Vol. 1, Marcel Dekker, NY, 1975, p. 1.
[16] Brochon, J. C. and Wahl, P., *European Journal on Biochemistry*, Vol. 25, 1972, pp. 20–32.
[17] Lipari, G. and Szabo, A., *Biophysical Journal*, Vol. 30, 1980, pp. 489–456.
[18] Tao, T., *Biopolymers*, Vol. 8, 1969, pp. 609–632.
[19] Belford, G. G., Belford, R. L., and Weber, G., *Proceedings of the National Academy of Science*, Vol. 69, 1972, pp. 1392–1393.
[20] Chuang, T. J. and Eisenthal, K. B., *Journal of Chemical Physics*, Vol. 57, 1972, p. 5094.
[21] Rigler, R. and Ehrenberg, M., *Quarterly Review of Biophysics*, Vol. 6, 1973, pp. 139–199.
[22] Wegener, W. A., Koester, V. J., and Dowben, R. M., *Proceedings of the National Academy of Science*, Vol. 76, 1979, pp. 6356–6360.
[23] Harvey, S. C. and Cheung, H. C., *Proceedings of the National Academy of Science*, Vol. 69, 1972, pp. 3670–3672.
[24] Harvey, S. and Cheung, H. C., *Biopolymers*, Vol. 19, 1980, p. 913.
[25] Cherry, R. J., *Biochimica Et Biophysica Acta*, Vol. 559, 1979, pp. 289–327.
[26] Kawato, S. K., Kinosita, K., Jr., and Ikegami, A., *Biochemistry*, Vol. 16, 1977, pp. 2319–2324.
[27] Kinosita, K., Jr., Kawato, S., and Ikegami, A., *Biophysical Journal*, Vol. 20, 1977, pp. 289–305.
[28] Lakowicz, J. R., Prendergast, F. G., and Hogan, D., *Biochemistry*, Vol. 18, 1979, pp. 508–519.
[29] Yguerabide, J., Epstein, H. F., and Stryer, L., *Journal of Molecular Biology*, Vol. 51, 1970, pp. 573–590.
[30] Hansen, D. C., Yguerabide, J., and Schumaker, V. N., *Biochemistry*, Vol. 20, 1981, pp. 6842–6852.
[31] Lakowicz J. R., and Prendergast, F. G., *Science*, Vol. 200, 1978, pp. 1399–1401.
[32] Weber, G., *Journal of Chemical Physics*, Vol. 66, 1977, pp. 4081–4091.
[33] Mantulin, W. W. and Weber, G., *Journal of Chemical Physics*, Vol. 66, 1977, pp. 4092–4099.
[34] Weber, G., Helgerson, S. L., Cramer, W. A., and Mitchell, G. W., *Biochemistry*, Vol. 15, 1976, pp. 4429–4432.
[35] Chen, R. F., *Journal of Biological Chemistry*, Vol. 242, 1967, pp. 173–181.
[36] Chen, R. F. and Bowman, R. L. *Science*, Vol. 147, 1965, pp. 729–732.
[37] Chen, R. F., *Archives of Biochemical Biophysics*, Vol. 133, 1969, pp. 263–276.
[38] Valeur, B. and Weber, G., *Photochemical Photobiology*, Vol. 25, 1977, pp. 441–444.
[39] Ross, J. B. A., Schmidt, C. J., and Brand, L., *Biochemistry*, Vol. 20, 1981, pp. 4369–4377.

Robert Weinberger,[1] Karen Rembish,[2] and L. J. Cline Love[3]

Comparison of Techniques for Generating Room Temperature Phosphorescence in Fluid Solution

REFERENCE: Weinberger, R., Rembish, K., and Cline Love, L. J., "**Comparison of Techniques for Generating Room Temperature Phosphorescence in Fluid Solution,**" *Advances in Luminescence Spectroscopy, ASTM STP 863*, L. J. Cline Love and D. Eastwood, Eds., American Society for Testing and Materials, Philadelphia, 1985, pp. 40–51.

ABSTRACT: Four techniques for inducing room temperature phosphorescence (RTP) in fluid solution are discussed, and the general requirements of molecular association for observation of RTP are presented. These methods are micelle-stabilized RTP, microcrystalline/colloidal RTP, cyclodextrin-induced RTP, and sensitized/quenched RTP. Examples of results using the different techniques for several classes of molecules, including carbocyclic and heterocyclic aromatic hydrocarbons and drugs, are used to illustrate the capabilities of these methods. The analytical utility of the four methods is compared and contrasted.

KEY WORDS: luminescence, phosphorescence, fluorescence, micelles, temperature, surfactants, colloids, drugs, heavy atoms, microcrystals, cyclodextrins, sensitized phosphorescence, quenched phosphorescence, polycyclic aromatic hydrocarbons

Observation of analytically useful phosphorescence was restricted, until recently, to solutes dissolved in or onto rigid matrices, often at cryogenic temperatures. This was necessary to minimize the quenching mechanisms that predominate in fluid solution because of the long lifetime of the triplet state. The transition from the triplet state to the ground state, whether radiative or radiationless, involves a change in the spin of the electron involved, and this transition is characterized as forbidden by quantum mechanical selection rules. The word "forbidden" if taken literally, provokes a feeling of something unusual by

---

[1]Senior applications scientist, Kratos Instruments, Ramsey, NJ.
[2]Fluorescence product manager, SPEX Industries, Inc., Metuchen, NJ.
[3]Professor of chemistry, Seton Hall University, Department of Chemistry, South Orange, NJ 07079.

modern day standards and something difficult to observe. On the other hand, these cases represent an opportunity for basic research, since the fruits of such investigations may lead to a better understanding of matter and, subsequently, an improvement of analytical methods.

The challenges of devising methods for the observation of room temperature phosphorescence (RTP) have led to schemes that alter the environment experienced by individual molecular species. This is accomplished in heterogeneous solvent matrices, such as aqueous solutions containing surfactants [1–6] or cyclodextrins [7,8], in colloidal suspensions [9,10] where the insoluble solute interacts with its crystalline neighbors, or in homogeneous solutions where donor-acceptor complexes may be of importance [11–13]. All of these mechanistically unique devices use some form of molecular association to stabilize the triplet state or permit the observation of RTP through energy transfer. This paper will describe and compare four methods of generating RTP in fluid solution, which are potentially useful in analytical chemistry. Each phosphorescence technique can be applied to many different classes of compounds and can yield drastically different spectroscopic information. The procedures to be discussed are micelle stabilized, colloidal/microcrystalline, and sensitized room temperature phosphorescence. A fourth technique, cyclodextrin RTP, the most recent and perhaps the most promising analytical RTP method, will be described briefly.

Experimental Procedures

Apparatus

All spectra were obtained using a Fluorolog® 2 + 2 spectrofluorometer (SPEX Industries, Metuchen, NJ) with double excitation and emission monochromators (spectral bandpass, 1.8 nm/mm), a 450-W xenon continuous light source, and a cooled Hammamatsu® R928 photomultiplier tube (PMT) operated in the photon counting mode. No temporal discrimination was used and all experiments were performed at ambient temperature. Data acquisition and manipulation were facilitated by a SPEX Datamate computer interfaced directly to the fluorometer. Hard copy printouts of spectra were produced by a Houston Instruments digital x-y recorder. All emission spectra reported have been corrected for lamp intensity and PMT response characteristics.

Reagents

All chemicals and concentrations used are contained in the figure captions.

Procedures

Specific experimental procedures have all been published and described in Ref. *1*, *5*, *6*, and *8* to *13*. The reader is referred to these papers if more detail is desired.

Results and Discussion

Micelle-Stabilized Room Temperature Phosphorescence (MS-RTP)

The vast array of chemical and biological literature on micelles is a testament to the importance of these microscopic aggregates in science and life itself. The properties of the heterogeneous micellar environment that provide uniqueness include solubilizing power (hydrophobic effects), electrostatic interactions, surface tension changes, and microviscosity changes [14]. Recently, analytical chemists have begun to exploit micellar solutions to perform chromatographic separations [15–17], enhance fluorescence [14] and chemiluminescence measurements [18], and to observe the phosphorescence of aromatic molecules at room temperature in fluid solution [1–6,16].

The normal micelle is comprised of surfactant molecules that aggregate above a certain critical micelle concentration to form roughly spherical clusters with the polar head group directed outward into the bulk solution and the tail pointed inward, forming a hydrocarbon-like pool. For sodium dodecyl sulfate (SDS), the surfactant used in these studies, it it possible to exchange the sodium counterions with thallium in a simple synthetic procedure [1]. A mixed micellar solution consisting of 30% thallium substituted surfactant (the solubility limit of mixed thallium/sodium SDS) is generally used for phosphorescence measurements of carbocyclic aromatic molecules. The properties of these heavy atom-substituted micellar solutions that result in phosphorescence from dissolved analytes are described below.

Exposure of Analyte to High Concentrations of Heavy Atoms — Hydrophobic molecules are attracted to the oil-like environment of the micelle and, in so doing, are placed in proximity of the heavy atom counterions whose local concentration is estimated to be 3 to 5 M. Thus, the analyte experiences an unusually high density of heavy atoms resulting in very efficient spin-orbit coupling, which can diminish the fluorescence and increase the phosphorescence. While it was orginally thought that this phenomenon greatly increased triplet state population, the predominate effects are more likely increased depopulation of the excited states. The fluorescence is diminished by increased radiationless depopulation and, in some cases, somewhat greater intersystem crossing rate from the excited singlet to the triplet state. The phosphorescence results from somewhat increased triplet state population via intersystem crossing and from appreciably increased radiative depopulation of the triplet state. For example, many aromatic molecules, such as naphthalene, have highly populated triplet state in fluid solution. However, in the absence of heavy atoms, no MS-RTP is observed, indicating loss by radiationless processes. Further evidence for these mechanisms comes from experiments with sensitized phosphorescence (to be described later), a technique that requires the analyte to have an appreciable, spontaneous population of its triplet state for subsequent energy transfer. Even in the absence of heavy atoms, strong sensitized signals are readily measured for many aromatic compounds, including naphthalene.

Minimizing Excimer Formation and Triplet-Triplet Annihilation — Triplet-triplet annihilation (TTA) is believed to be one of the priniciple reasons that phosphorescence is not generally observed for many aromatic hydrocarbons in homogeneous fluid solution. In micellar media, if the solute concentration is low, segregation of the analytes occurs, such that the micelle occupancy is no more than one analyte molecule per micelle. Indeed, using Poisson-Boltzman statistics, at low analyte concentrations ($<1 \times 10^{-5} M$), greater than 95% of the micelles contain no analyte molecules. For a surfactant with an aggregation number (number of associated surfactant monomers) of 60 and a surfactant concentration of $0.15 M$, the percentage of multiple-analyte occupancy for a $2.5 \times 10^{-4} M$ solution is only 3.4% [1]. This also explains why other concentration dependent phenomena, such as excimer formation, which diminishes the molecular triplet emission intensity, is not observed.

Protection from Quenchers — Arguments can be made for some measure of protection of analytes from quenchers by the micellar assembly. However, the dynamic nature of micelle formation and the dynamics of solute association/dissociation with the micelle prevents effective shielding of the solute from quenchers. Any quencher that can permeate the micelle will generally quench the associated analyte; however, if the quencher is excluded from the micelle, such as by electrostatic repulsion of anionic quenchers from anionic micelles, some protection is afforded. Much better protection from quenchers is achieved in more rigid or stable matrices, such as with microcrystalline dispersions or cyclodextrin solutions. All MS-RTP spectra must be obtained in oxygen-free solutions to prevent quenching of the triplet state by dissolved oxygen, as oxygen easily penetrates the micelle. Complete or partial freedom from oxygen quenching is observed using the other two techniques mentioned above, which use more organized media.

The excitation, residual fluorescence, and MS-RTP spectra for biphenyl is shown in Fig. 1. The phosphorescence spectrum is slightly red shifted with considerable loss of fine structure compared to its 77 K spectrum [19] (not shown). The red shift suggesting triplet state stabilization can be partially accounted for by Franck-Condon reorganization of the excited molecule, possible only in the fluid environment. Loss of fine structure is indirect evidence for increased vibronic congestion, which is supported by data obtained in microcrystalline matrices where the molecules are much more firmly locked in position.

Intense MS-RTP is observed for most two- to five-ring polynuclear aromatic hydrocarbons, biphenyls, and some of their derivatives. Intense emissions are observed for heterocyclic fluorene derivatives, such as dibenzofuran and dibenzothiophene [20]. Nitrogen heterocycles, such as quinoline and phenazine, give weaker emission, which is only observable using silver heavy atom counterion micelles [21]. There is strong evidence that complexation of silver with the nitrogen electrons or the π electron system is necessary to decrease the exit rates of these more water soluble species from the micelle. When outside the micellar environment in the bulk aqueous solvent, it is unlikely that RTP can be observed

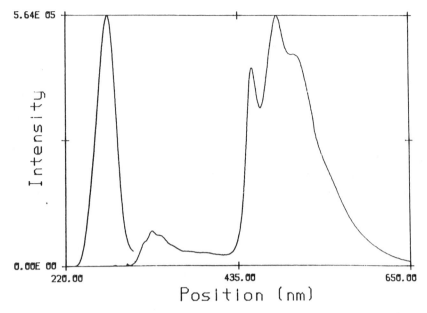

FIG. 1—*Excitation, corrected fluorescence and corrected phosphorescence spectra of 1 × 10^{-4} M biphenyl in 0.1 M Tl/Na dodecyl sulfate (30:70); excitation wavelength = 273 nm; slits: 5.8-nm excitation and 2.1-nm emission.*

because of the reasons outlined above. No MS-RTP is observed from more flexible molecules such as dipenzopdioxin, benzophenone, or diphenylmethane. These molecules lack the structural rigidity that is one of the prerequisites for observation of MS-RTP. Phosphorescence has been observed from pharmaceutically important compounds, for example, propranolol and its 4-hydroxy metabolite, diflunisal, naproxen, and naphazoline [20]. It is possible to spectroscopically separate propranolol from its metabolite with synchronous wavelength scanning second derivative MS-RTP [22].

Limits of detection by MS-RTP can approach 1×10^{-10} M with strong emitters [6] and is usually restricted by impurities that are present in the reagents. Using MS-RTP liquid chromatographic detection, low nanogram sensitivities have been recorded [16]. This approach is limited by problems in deaerating litre quantities of the foaming, bubbling surfactant solutions used as the mobile phase. Adaptation of RTP detection to microbore liquid chromatography may solve this problem, since degassing a 50-mL solvent reservoir is more easily accomplished.

Microcrystalline Phosphorescence

Intense room temperature phosphorescence has been observed from aqueous microcrystalline suspensions of polycyclic aromatic hydrocarbons (PAHs) [9,10]. These suspensions are prepared by adding the solute as a concentrate dissolved in a water miscible solvent. The stability of the preparation is surprising; in glass containers, they can remain suspended for at least several weeks. Observation under a light polarizing microscope reveals substantial Brownian motion, accounting partially for their stability. A significant Tyndall effect was observed also. Anthracene microcrystals were needle-like, with a maximum size of less than 100 μm in length. Pyrene crystals were smaller and their shape not readily discernible. These suspensions meet the criteria for colloids except that the mean particle size is larger than generally recognized for such a classification [9,10,20].

The spectra of biphenyl, dissolved in methanol and suspended in water, are shown in Fig. 2. In the suspension, four peaks that are clearly due to phosphorescence emission appear on the tail of the fluorescence. The shape, symmetry, and resolution of these peaks are remarkably similar to that obtained at 77 K in EPA solvent, although a 77 K hexane solvent provides better resolution of the vibronic fine structure [19]. The wavelengths of emission from the suspension are blue shifted by 50 nm (0-0 bands) relative to those from the cryogenic matrix.

The rationale for the appearance of RTP is quite reasonable. In a manner analogous to solid-substrate RTP, the quenching mechanisms that operate in fluid solution are minimized. In this context, microcrystals serve as their own substrate. Since the solute is present as a solid in suspension, the RTP was expected

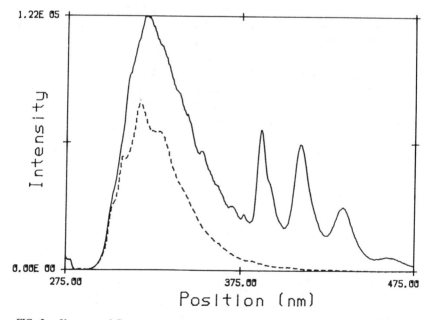

FIG. 2—*Uncorrected fluorescence and phosphorescence spectra of 50 μg/ml of biphenyl dissolved in methanol (dashed line) and suspended in water (solid line); excitation wavelengths: in methanol 278 nm, and in water 271 nm; slits = 0.9-nm emission, 7.2 and 5-nm excitation for aqueous and methanol solutions, respectively. Note the appearance of phosphorescence above 350 nm in the spectrum of the aqueous solution (0.5-s integration/point).*

and was found to be immune to oxygen quenching. The lifetime of the biphenyl emission was found to be 70 μs [20], much shorter than the 400 μs measured using the MS-RTP technique.

The wavelengths of emission for all compounds measured were blue shifted relative to MS-RTP and are shown in Table 1. Microcrystalline RTP has been observed for diverse structures such as biphenyls, PAHs, heteroatom derivatives of fluorene, and a few drugs. No RTP was found for 1,1-binaphthyl, p-terphenyl, p-quaterphehyl, dibenzopdioxin, carbazole, chrysene, pyrene, corenene, and triphenylene. However, through delayed fluorescence emission, the triplet state plays an important role in the microscrystalline luminescence for many of these compounds [20].

Other than RTP, important spectral changes may also be noted in microcrystalline suspensions. Compounds with no carbon sharing more than two aromatic rings form Type A crystal lattices and generally exhibit strong self-absorption whenever there is overlap of the 0-0 absorption and 0-0 fluorescence bands. Anthrancene and chrysene are typical examples of compounds undergoing this effect. These features, of course, can be generated in highly concentrated fluid solutions. In the suspension, self-absorption is measurable at much lower concentrations as long as the concentration is higher than the solubility limit of the solute [9].

Type B crystals, where a carbon can be shared by three aromatic rings, produce an excimer-like emission in the microcrystalline suspensions. In these lattices, the molecular overlap is sufficient for interaction between adjacent molecules, which is necessary since molecular diffusion is nonexistent in the solid state. Pyrene, chrysene, and coronene all exhibit this phenomenon. However, this emission must be termed "excimer-like," since the excitation spectra show that an interaction between the adjacent molecules is also occurring in the ground state. Properly speaking, excimer interaction is defined as an excited state

TABLE 1 — *Comparison of room temperature phosphorescence wavelengths from microcrystalline suspensions and micellar solutions.*

	Phosphorescence Wavelengths, nm[a]		
Compound	Microcrystals	MS-RTP	$\Delta\lambda$ (0-0)[b]
Biphenyl	386, 409, 434, 461	452, 482, 505	66
4-Bromobiphenyl	376, 389, 410, 442	498, 530, 560	122
2-Bromonaphthalene	394, 414, 440	495, 525, 563, 613(s)	101
Dibenzofuran	400, 422, 444, 471	424, 450, 480(s), 516(s)	24
Fluorene	330, 346, 362, 383	437, 465, 491, 533(s)	107
Phenanthrene	402, 412, 447, 480	464, 516, 553, 609(s)	62
Naphthalene	387, 410, 436, 462(s)	486, 521, 557, 601	99

[a](s) denotes shoulder on peak.
[b]$\Delta\lambda$ is wavelength difference of 0-0 vibronic bands from the two techniques.

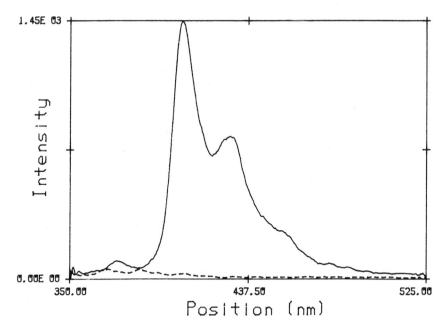

FIG. 3—*Uncorrected delayed fluorescence spectrum of 10-µg/mL chrysene dissolved in methanol* (dashed line) *and suspended in water* (solid line); *excitation wavelengths: 275 nm for both; time delay = 30 µs; and emission slits = 7.2 nm.*

phenomenon. In the Type A crystal, the excitation spectra are almost identical in suspension and in solution; this is not the case for Type B crystals [9].

For many of the compounds not exhibiting RTP, a delayed component of the emission, whether excimer or molecular in origin, was observed under conditions of temporal discrimination. This is illustrated for chrysene (Fig. 3) where, with a 30-µs time delay between excitation and measurement of emission, appreciable emission is observed only for the suspended material. The excimer-forming compounds like pyrene and coronene contain similar delayed components in their emissions [20].

All of these luminescence properties of the microcrystalline suspensions are dependent on the solubility of each solute in water. It seems quite reasonable that this technique might be useful in determining the solubility of these rather insoluble compounds in water. Accordingly, it may be possible to determine the state of a compound in a particular matrix, without specimen preparation. This may prove to be important since it has been shown that many potent carcinogens are more toxic in the solid state [23].

Sensitized Room Temperature Phosphorescence

Triplet state energy transfer in fluid solution is a well documented and a relatively common phenomenon (24–27). The process involves energy transfer from a donor molecule to the triplet state of an acceptor. If the acceptor molecule then emits phosphorescence radiation, the donor is said to sensitize the emission of the acceptor. Recently, Donkerbroek and co-workers have employed the sensitization phenomena to develop a new detector for liquid chromatography based on RTP without the use of micelles to stabilize the triplet state [11–13]. The recent thrust of their work has been with biacetyl as the acceptor molecule, a species that is certainly an anomoly in photophysics. It is the only aliphatic molecule known to phosphoresce in fluid solution and has been termed "pathologic" by Turro. Nevertheless, the sensitized biacetyl phosphorescence in homogeneous fluid solution is intense enough to be analytically useful. The fundamental requirements for an analytical method based on sensitized phosphorescence are the following:

1. The triplet energy of the donor (analyte) is greater than that of the acceptor.
2. The donor triplet state lifetime is greater than the time for energy transfer.
3. The donor triplet state is appreciably populated.
4. The donor can be excited at a wavelength where the acceptor does not absorb.

Compounds, such as naphthalenes, biphenyls, dibenzofurans, and benzophenones, have been shown capable of sensitizing biacetyl emission. The limits of detection in a static mode are as low as $1 \times 10^{-9} M$, and the picogram level can be measured in a liquid chromatographic flow system [13]. More recently, pharmaceutical agents, such as cocaine, brethine, didrate, estradiol, methaqualone, sulfanilamide, and phenobarbital, have been shown to be strong sensitizers [20]. However, the majority of drugs that met requirements 1, 3, and 4 did not sensitize biacetyl emission, presumably because of short triplet lifetimes or poor spectral overlap integrals. The spectrum of biacetyl, as sensitized by p-dibenzofuran, is shown in Fig. 4.

Sensitized phosphorescence represents a viable alternative to micelles as a means of observing molecules, albeit indirectly, via their triplet state. A significant advantage of the technique over MS-RTP is the ability to employ solvents commonly used in liquid chromatography as the energy transfer medium. The means that well-characterized reverse phase separations are suitable for chromatography using sensitized RTP detection, and no new chromatography need be developed. One disadvantage is that most of the qualitative information from the analysis is lost, since biacetyl emission is always being measured. This can be used advantageously in liquid chromatography since the emission monochrometer of the luminescence detector can be set at a constant wavelength for all analytes. The apparent loss of selectivity is not critical because the phosphorescence assay itself is quite selective. The latest advance in this field is based on quenching of biacetyl emission by analytes [28]. If biacetyl is radiationally excited, it is possible to measure the quenching of this emission by both organic and inorganic species. When a quencher elutes from a liquid chromatographic column, a signal will be detected that is a diminished biacetyl emission. This may

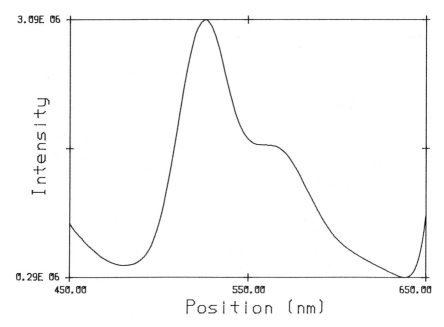

FIG. 4—*Phosphorescence spectrum of 1×10^{-4} M biacetyl, as sensitized by 1×10^{-4} M p-dibenzofuran in 0.1 M Tl/Na dodecyl sulfate; excitation wavelength = 295 nm; slits: 5.8-nm excitation and 1-nm emission; 0.5-s integration/point.*

well be the most versatile phosphorescence technique since many species are effective quenchers of phosphorescence.

Cyclodextrin Room Temperature Phosphorescence (CD-RTP)

The latest advance in RTP involves another form of ordered media, the cyclodextrins (CDs). These molecules are macrocyclic carbohydrates that form a doughnut-like structure, the size of the doughnut "hole" being determined by the number of units in the ring [29]. CDs form inclusion complexes with many molecules in fluid solution if the steric properties of the molecule permit it to at least partially enter the CD cavity. If a heavy atom reagent, such as dibromoethane, is also present in solution, intense RTP signals are induced in the trimolecular complex composed of CD:analyte:heavy atom species [8,30]. The stable, protective nature of most of the inclusion complexes is impressive in that the CD provides partial protection of the triplet state from oxygen quenching, a feature not seen, to date, in micellar solutions. Limits of detection have approached 1×10^{-13} M, making it the most sensitive of the RTP techniques [8]. The spatial constraints of the CD cavity appear to increase the vibronic structure of the complexed analyte, as evidenced by enhanced spectral resolution of CD-RTP compared to MS-RTP. However, the resolution is less than obtained with the microcrystalline technique. Exciting developments are expected with cyclodextrin-induced RTP in the near future.

Molecular Association

The common theme expressed by all of the RTP techniques is the necessity of some form of molecular association. In a manner analogous to solid substrate RTP or low-temperature phosphorescence, three of the four techniques discussed utilize molecular organization to stabilize the phosphorescence emission of the individual solute molecules, and the fourth technique uses molecular association of the donor-acceptor molecules. These techniques differ from the more classical methods by the degree of dynamic interaction with the microenvironment.

Microcrystalline RTP most closely resembles the immobilization techniques since the matrix is fixed at the time of preparation. The spectral resolution is indicative of the strongly bound nature of the microcrystalline environment. Cyclodextrin RTP results from a more dynamic system compared to the microcrystal since the solute has the opportunity to exit the cavity (although with a much lower exit rate than entrance rate) and is less vibrationally constrained. The micellar solution represents the most dynamic technique. The spectra are less resolved than the previous two, and the exit and entrance rates of solutes are rapid. All three of these techniques provide some form of immobilization, protection, or both, which serves to permit the observation of RTP.

For sensitized phosphorescence, dynamic immobilization is unnecessary since the lumiphor is one of the rare molecules that readily phosphoresces in homogeneous fluid solution. However, dynamic, diffusion controlled molecular organization plays a crucial role in the formation of the donor-acceptor complex. Factors that can improve the degree of donor-acceptor complex formation can enhance the sensitivity of the method, for example, the selection of solvents providing enhanced energy transfer. It has been shown that the addition of micelles can improve the sensitivity of sensitized phosphorescence, presumably because of drawing both donor and acceptor into close proximity to one another, providing a higher probability of energy transfer compared to the strictly diffusion limited process [20]. It would be expected that other types of ordered media, for example, liquid crystals, could profoundly influence many forms of luminescence spectroscopy. If other types are to be analytically useful, experimental procedures must be devised to permit facile production of the desired spectroscopic observable with minimum specimen manipulation.

Acknowledgments

The authors are grateful to Stephen Scypinski for providing data on cyclodextrin-induced room temperature phosphorescence, and to Roland Frei, Nel Velthorst, and Jan Donkerbroek for helpful discussion of both sensitized and quenched room temperature phosphorescence. This work was supported in part by the National Science Foundation Grant CHE-8216878 and the Environmental Protection Agency. Although the research described in this article has been funded by the U.S. Environmental Protection Agency under assistance agreement R809474 to L. J. Cline Love, it has not been subjected to the Agency's required peer and administrative review and therefore does not necessarily reflect the view of the Agency, and no official endorsement should be inferred.

References

[1] Cline Love, L. J., Skrilec, M., and Habarta, J., *Analytical Chemistry,* Vol. 52, 1980, pp. 754–759.
[2] Kalyanasundaram, K., Grieser, F., and Thomas, J. K., *Chemical Physics Letters,* Vol. 51, 1977, pp. 501–505.
[3] Turro, N. J. and Aikawa, M. J., *Journal of the American Chemical Society,* Vol. 102, 1980, pp. 4866–4870.
[4] Humphry-Baker, R., Moroi, Y., and Gratzel, M., *Chemical Physics Letters,* Vol. 58, 1978, pp. 207–210.
[5] Skrilec, M., Cline Love, L. J., *Journal of Physical Chemistry,* Vol. 85, 1981, pp. 2047–2050.
[6] Skrilec, M. and Cline Love, L. J., *Analytical Chemistry,* Vol. 52, 1980, pp. 1559–1562.
[7] Turro, N. J., Cox, G. S., and Li, X., *Photochemistry and Photobiology,* Vol. 37, 1983, p. 149.
[8] Scypinski, S. and Cline Love, L. J., *Analytical Chemistry,* Vol. 56, 1984, pp. 322–327.
[9] Weinberger, R. and Cline Love, L. J., *Spectrochima Acta, Part A, Molecular Spectroscopy,* Vol. 40, 1984, pp. 49–55.
[10] Weinberger, R. and Cline Love, L. J., submitted for publication in *Applied Spectroscopy,* in press.
[11] Donkerbroek, J. J., Elzas, J. J., Goojier, C., Frei, R. W., and Velthorst, N. H., *Talanta,* Vol. 28, 1981, p. 717.
[12] Donkerbroek, J. J., Goojier, C., Velthorst, N. H., and Frei, R. W., *Analytical Chemistry,* Vol. 54, 1982, pp. 891–895.
[13] Donkerbroek, J. J., van Eikema Holmes, N. J. R., Goojier, C., Velthorst, N. H., and Frei, R. W., *Chromatographia,* Vol. 15, 1983, p. 218.
[14] Hinze, W. L., *Solution Chemistry of Surfactants,* Vol. I, K. L. Mittal, Ed., Plenum Press, New York, 1979, pp. 79–127.
[15] Yarmchuk, P., Weinberger, R., Hirsch, R. F., and Cline Love, L. J., *Analytical Chemistry,* Vol. 54, 1982, pp. 2233–2238.
[16] Weinberger, R., Yarmchuk, P., and Cline Love, L. J., *Analytical Chemistry,* Vol. 54, 1982, pp. 1552–1558.
[17] Armstrong, D. W. and Nome, F., *Analytical Chemistry,* Vol. 53, 1981, pp. 1662–1666.
[18] Hinze, W. L., *Abstract of Papers of the 9th FACSS Meeting,* Abstract 514, FACSS, Philadelphia, Sept. 1982.
[19] Kanda, Y., Ihimda, R., and Sakai, Y., *Spectrochima Acta,* Vol. 17, 1961, pp. 1–6.
[20] Weinberger, R., Ph.D. dissertation, Seton Hall University, South Orange, NJ, 1983.
[21] Woods, R. and Cline Love, L. J., *Spectrochima Acta, Part A, Molecular Spectroscopy,* Vol. 40, 1984, pp. 643–650.
[22] Femia, R. and Cline Love, L. J., *Analytical Chemistry,* Vol. 56, 1984, pp. 327–331.
[23] Lakowicz, J. R., Englund, F., and Hedmork, A., *Biochemistry Biophysics Acta,* Vol. 543, 1978, pp. 202–206.
[24] Backstrom, H. L. J. and Sandros, K., *Acta Chimica Scandinavica,* Vol. 12, 1958, p. 823.
[25] Richtol, H. H. and Klappmeier, F. H. J., *Chemical Physics,* Vol. 44, No. 4, 1966, p. 1519.
[26] Almgren, M., *Photochemistry and Photobiology,* Vol. 6, 1967, p. 829.
[27] Turro, N. J., Chiang, L. K., Ming-Fea, C., and Lee, P., *Photochemistry and Photobiology,* Vol. 26, 1969, p. 523.
[28] Donkerbroek, J. J., van Eikema Holmes, N. J. R., Goojier, C., Velthorst, N. H., and Frei, R. W. J., *Chromatogr.,* Vol. 255, 1982, pp. 581–590.
[29] Bender, M. L. and Komiyama, M., *Cyclodextrin Chemistry,* Springer-Verlag, New York, 1979, Chapters II and III.
[30] Scypinski, S. and Cline Love, L. J., *Analytical Chemistry,* Vol. 56, 1984, pp. 331–336.

Coupled Phenomena in Luminescence

Gordon F. Kirkbright[1]

Some Photoacoustic Studies of Consequence in Luminescence Spectroscopy

REFERENCE: Kirkbright, G. F., "**Some Photoacoustic Studies of Consequence in Luminescence Spectroscopy,**" *Advances in Luminescence Spectroscopy, ASTM STP 863*, L. J. Cline Love and D. Eastwood, Eds., American Society for Testing and Materials, Philadelphia, 1985, pp. 55–62.

ABSTRACT: In photoacoustic spectroscopy (PAS) the signal response with respect to wavelength of incident radiation is detected as heat at the surface of the sample that results from nonradiative relaxation of the excited states produced on absorption by the molecular species in the sample. This process is complementary to radiative relaxation via luminescence, and study of photoacoustic signals can give information concerned with the quantum efficiency of luminescence processes. The manner in which absolute quantum efficiencies can be determined for both solid and liquid samples with simple instrumentation and without reference to luminescence standard materials is described.

KEY WORDS: spectroscopy, acoustics, radiation, quantum efficiency, fluorescence

In recent years there has been a rapid rise of interest in the photoacoustic effect and its application in the examination of solid and liquid samples. Harshbarger and Robin [1], Rosencwaig [2], and Adams et al [3,4] have demonstrated the technique of analytical photoacoustic spectrometry (PAS) in studies of a variety of sample types.

In PAS, intensity modulated electromagnetic radiation is incident upon the sample material enclosed in a cell of constant volume. If the sample absorbs at the wavelength of the incident radiation, on subsequent deexcitation the absorbed energy may appear as heat and cause a periodic pressure rise in the gas surrounding the sample. This change in pressure may be detected and monitored by a sensitive microphone transducer enclosed within the cell. The resulting electrical signal is selectively amplified using a tuned amplifier and phase-sensitive

---

[1]Deceased, see dedication; formerly, professor of analytical science, Department of Instrumentation and Analytical Science, University of Manchester Institute of Science and Technology, Manchester, United Kingdom.

detector and presented as a DC potential to a potentiometric chart recorder or other voltage display system.

Clearly, PAS relies on the radiationless conversion of absorbed energy for the production of an acoustic signal and, hence, is complementary to the conventional techniques of luminescence spectrometry. As a pressure transducer is employed to monitor indirectly the temperature of the absorbing species, photoacoustic spectrometry may be considered as a calorimetric method for the detection and study of excited states. Rosencwaig [2] suggested that the technique might be useful for these purposes but presented no data relating to liquid samples. Callis [5] has reviewed some modern calorimetric techniques for examining radiationless deactivation in solid and liquid samples employing both microphones and piezoelectric crystals as heat-flow transducers. More recently, Lahmann and Ludewig [6] have employed the photoacoustic technique using a piezoelectric transducer immersed in the sample solution to determine the absolute fluorescence quantum efficiency of rhodamine 6G in aqueous solution. Such calorimetric methods of studying excited states are reported to be rapid, simple, and sensitive.

PAS has been used for the determination of the absolute quantum efficiency of quinine bisulfate in aqueous solution and examination of the quenching effect on this species observed upon the addition of chloride ions to the sample solution. In addition these studies have been extended to the examination of solid sample materials.

Theory

To estimate the absolute quantum efficiency of fluorescence Q for a compound from PAS data, it is necessary to derive a theoretical expression for the magnitude of the photoacoustic signal. For constant experimental conditions and assuming no optical or thermal saturation

$$P(\text{PAS}) = P_{abs} \cdot \gamma \tag{1}$$

where $P(\text{PAS})$ is the magnitude of the PAS signal at wavelength λ, P_{abs} is the radiant power absorbed at this wavelength by the sample, and γ is an efficiency factor that is a measure of the conversion efficiency of absorbed power into heat by nonradiative mechanisms.

For a sample capable of fluorescence it can be shown that

$$\gamma = 1 - Q + Q[(\nu_o - \nu_F)/\nu_o] \tag{2}$$

where Q is the quantum efficiency, ν_o is the frequency of the exciting radiation, and ν_F is the mean frequency of the fluorescent radiation.

In Eq 2, the term $(1 - Q)$ accounts for the absorption of radiation by nonfluorescing molecules, and the second term corrects for the radiationless fraction of energy produced in the sample in the fluorescence process (that is, the Stokes shift).

Thus, combining Eqs 1 and 2, for a fluorescent material

$$P(\text{PAS})_F = P_{abs}\{1 - Q + Q[(\nu_o - \nu_F)/\nu_o]\} \tag{3}$$

and for a material for which the nonradiative relaxation time is shorter than the radiative relaxation time (that is, a nonfluorescent material), where $\gamma \approx 1$, we have

$$P(\text{PAS})_{NF} = P_{abs} \tag{4}$$

Assuming the absorption coefficients of both the fluorescent and nonfluorescent compounds to be identical and the thermal characteristics of the solutions to be similar, P_{abs} is constant for both samples and Eq 3 may be expressed upon rearrangement as

$$Q = \lambda_F/\lambda_o[1 - P(\text{PAS})_F/P(\text{PAS})_{NF}] \tag{5}$$

where λ_F is the mean wavelength of fluorescence and λ_o the wavelength of excitation.

Previous studies concerning the determination of absolute quantum efficiencies of fluorescent materials utilizing PAS have employed pressure transducers immersed in the analyte and nonfluorescent materials as reference solutions [5,6]. This is achieved by ensuring that the reference material and the fluorescent compound to be examined have comparable absorption coefficients and by equating the difference in magnitude of the PAS signals from the two solutions to the efficiency of fluorescence of the sample. The method described here in a study of quinine bisulfate uses this compound as both the sample and the reference. This is achieved by monitoring the PAS signals from the fluorescent quinine sulfate solution and subsequently the signals obtained from nonfluorescent sample solutions after addition of chloride ions to promote quenching.

Halide ions are efficient fluorescence quenching agents, and there are several references in the literature pertaining to the total quenching of quinine bisulfate aqueous solutions in the presence of excess chloride ions.

To ensure that the addition of chloride ions to the quinine bisulfate solutions did not affect the absorptivity of the solutions in the concentration range of interest, absorbance versus analyte concentration plots were obtained using the photoacoustic spectrometer in the transmittance mode. The plots were obtained linear in accordance with Beer's law and of equal molar absorptivity irrespective of whether the quinine bisulfate is present in sulfuric acid (that is, fluorescent) or hydrochloric acid (quenched).

By this technique, a value for Q for quinine bisulfate (in 0.1 N sulfuric acid) over the concentration range 10^{-3} to 10^{-2} M was found to be 0.53 ± 0.02. This value is in good agreement with the accepted literature value of 0.51 [7].

The problems associated with the study and determination of photoluminescence quantum efficiency values of solid materials have always been more difficult to overcome than for materials in solution. Tregellas-Williams [8] has reviewed the methods employed for the determination of luminescence efficiencies for inorganic phosphors, and more recently, Lipsett has provided an extensive summary of the subject [9]. The majority of techniques employed today use photometric methods, that is, the detection and measurement of a certain fraction of the emitted luminescence following excitation of the sample, and an examination of the literature suggests that the most serious errors in these methods occur in applying corrections necessary to account for the experimental geometry used. Wrighton et al [10] have utilized a conventional scanning emission spectrophotometer to determine absolute fluorescence efficiencies of powdered samples. By measuring the diffuse reflectance of the sample relative to that of a nonabsorbing standard material at the excitation wavelength and by recording the emission of the sample under identical conditions, the luminescence efficiency was determined as the ratio of the intensity of emitted radiation to the difference in intensity of the diffuse radiation from the sample and the nonabsorbing standard. Following calibration of the detector sensitivity as a function of wavelength, however, and the introduction of corrections for the nonideality of the absolute reflectance standards, the error in the method was reported to be ±25%.

Considering the extensive application of photometric methods and their inherent disadvantages, there have been few developments in alternative techniques for determining luminescence quantum efficiencies of solids. Most calorimetric methods are modified versions of Bodo's and employ thermocouples or thermistors to monitor the heating effect within the sample following absorption of electromagnetic radiation. A piezoelectric calorimeter has been described by Callis [5] and used to determine the triplet yield for anthracene dissolved in a rigid matrix of polymethylmethacrylate. We have now developed the use of the technique of PAS in a constant volume gas cell for the examination of luminescence quantum efficiencies of solid materials. By utilizing the phenomenon of photoacoustic signal saturation, we wish to demonstrate that photoacoustic spectroscopy may provide a relatively rapid and precise method for the study of luminescence in solid materials. Luminescence quantum efficiency data are determined here for 1,1,4,4-tetraphenyl-1,1,3-butadiene, 2,2'-dihydroxyl-1, 1'-naphthaldiazine, and sodium salicylate.

Theoretical Considerations

For a sample capable of fluorescence, then as in Eq 2

$$\gamma = 1 - Q + Q[(\nu_o - \nu_F)/\nu_o]$$

where Q is the fluorescence quantum efficiency, ν_o is the frequency of excitation, and ν_F is the mean frequency of the fluoresced radiation. Combining Eqs 1 and 2 gives

$$P_{PAS}^F = K_F P_{abs(F)}\{1 - Q + Q[(\nu_o - \nu_F)/\nu_o]\} \qquad (6)$$

for a fluorescent material, and

$$P_{PAS}^{NF} = K_{NF} P_{abs(NF)} \qquad (7)$$

for a nonfluorescent sample, that is, $\gamma = 1$. The subscripts F and NF refer to the fluorescent and nonfluorescent conditions, respectively, and K is a constant given by the thermal transfer characteristics of the sample and the instrumental arrangement employed.

Thus, combining Eqs 6 and 7 and rearranging, in terms of wavelength

$$Q = \lambda_F/\lambda_o \left[1 - \frac{P_{PAS}^{F}}{P_{PAS}^{NF}} \frac{K_{NF}}{K_F} \frac{P_{abs(NF)}}{P_{abs(F)}} \right] \qquad (8)$$

In our earlier studies examining aqueous solutions of quinine bisulfate, the fluorescence was effectively quenched by the addition of chloride ions, without changing the absorption spectrum of the sample. Thus, the thermal and optical absorption characteristics of the fluorescent sample and nonfluorescent reference were identical and Eq 5 results; thus Q may then be readily determined.

Unfortunately, for solid samples the accurate matching of thermal and optical characteristics of the sample with a nonluminescent reference material is not easily achieved, and it is necessary to develop a new approach. This may be attained by utilizing the phenomenon of photoacoustic signal saturation. Rearrangement of Eq 8 provides

$$P_{PAS}^{F}/P_{PAS}^{NF} = (K_F/K_{NF})[P_{abs(F)}/P_{abs(NF)}][1 - (\lambda_o/\lambda_F)Q] \qquad (9)$$

which predicts a linear relationship between P_{PAS}^{F}/P_{PAS}^{NF} and λ_o, the excitation wavelength.

Rosencwaig and Gersho [11] and McClelland and Kniseley [12] have defined the thermal depth L of a photoacoustic signal as that thickness of sample contributing to the production of the photoacoustic signal at the sample-gas interface. The thermal depth is controlled by the thermal diffusivity, α cm^{2}s^{-1}, of the sample and the instrumental modulation frequency, ω rad s^{-1}, viz

$$L = (2\alpha/\omega)^{1/2} \qquad (10)$$

McClelland and Kniseley [12] have demonstrated that the photoacoustic signal is directly proportional to the sample absorption coefficient β for values at $\beta < 2\pi/L$ but becomes progressively less sensitive to increase in β, ultimately becoming independent of β at $\beta \simeq 20\pi/L$. The photoacoustic signal is then said to be "saturated." Many solid materials and powders have high absorption coefficients owing to their highly condensed form, and it may be anticipated that the photoacoustic signals derived from such samples will be independent of their absorption coefficients. Therefore, the serious problems of absorbance matching of samples and reference materials may be avoided by making use of photoacoustic saturation. Thus, if the photoacoustic signals are monitored in saturation and with identical instrumental conditions

$$P_{abs(F)} = P_{abs(NF)} \tag{11}$$

and Eq 9 reduces to

$$P^F_{PAS}/P^{NF}_{PAS} = (K_F/K_{NF})[1 - (\lambda_o/\lambda_F)Q] \tag{12}$$

A plot of P^F_{PAS}/P^{NF}_{PAS} versus λ_o (that is, the normalized photoacoustic spectrum) should yield a straight line of slope m and an intercept, at $\lambda_o = 0$, of K_F/K_{NF} where

$$m = (-K_F/K_{NF})(Q/\lambda_F) \tag{13}$$

The quantum efficiency of luminescence Q may be calculated from Eq 13. It is obviously necessary to ensure that as far as possible $K_F = K_{NF}$, and these conditions can be ensured using careful sample preparation techniques in which thin layers of samples are deposited onto thick substrates [13].

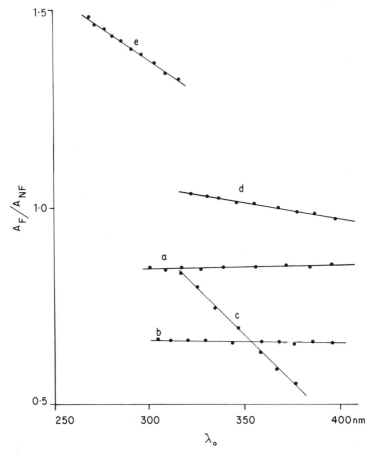

FIG. 1—*Normalized photoacoustic spectra within the wavelength range for photoacoustic saturation of (a) Congo red, (b) malachite green, (c) TPB, (d) yellow liumogen, and (e) sodium salicylate.*

TABLE 1—*Luminescence quantum efficiencies of materials examined.*

Sample Material	Luminescence Quantum Efficiency
TPB[a] (99% purity)	0.94 ± 0.01
TPB[a] (scintillation grade)	0.86 ± 0.01
	0.86 ± 0.01
Yellow Liumogen	0.33 ± 0.01
Sodium salicylate	0.55 ± 0.04
Quinine bisulfate (solution)	0.53 ± 0.02

[a]TPB is 1,1,4,4 tetraphenyl-1,1,3 butadiene.

TABLE 2—*Luminescence quantum efficiencies for some substituted benzthiazoles by conventional and photoacoustic spectroscopic methods.*

R_3	R_4	R_5	R_6	Conventional ϕ_f	PAS ϕ_f
H	H	H	H	0.38	0.44
CH$_3$	H	H	H	0.36	0.60
H	CH$_3$	H	H	0.39	0.46
H	H	CH$_3$	H	0.35	0.42
CH$_3$	H	CH$_3$	H	0.23	0.27
Cl	H	H	H	0.47	0.54
H	Cl	H	H	0.33	0.40
H	H	Cl	H	0.25	0.34
Cl	H	Cl	H	0.37	0.51
Br	H	H	H	0.25	0.35
H	Br	H	H	0.21	0.30
H	H	Br	H	0.27	0.26
Br	H	Br	H	0.32	0.41
OH	H	H	H	0.066	...
H	OH	H	H	0.39	0.40
H	H	OH	H	0.25	0.27
H	H	H	OH	0.018	...
OCH$_3$	H	H	H	0.22	0.20
H	OCH$_3$	H	H	0.46	0.53
H	H	OCH$_3$	H	0.07	...
NO$_2$	H	H	H	0.05	0.28
H	H	NO$_2$	H	0.02	...
NO$_2$	H	NO$_2$	H	0.009	...
H	H	NH$_2$	H	0.04	...
H	H	NHCOCH$_3$	H	0.18	0.28
H	H	NHCOPh	H	0.25	0.43
H	H	N = CHPh	H	0.16	0.19

Results

The technique described above has been applied to the determination of the luminescence quantum efficiency of various dye stuffs. Typical plots of P_{PAS}^{F}/P_{PAS}^{NF} versus λ_o are shown in Fig. 1, and the corresponding values of Q obtained from the slope of these graphs listed in Table 1. A considerable volume of work has recently been undertaken to investigate the effect on luminescence quantum efficiency of different substituents in a series of benzthiazole compounds; typical results are shown in Table 2. The PAS results are compared here with those obtained using the diffuse reflectance spectroscopy technique.

Conclusion

The versatility of PAS in the study of luminescence in condensed matter promises an excellent future for the technique. A method has been developed for the accurate determination of the absolute luminescence quantum efficiency of solid materials, which are otherwise difficult to study. The only criterion for the application of the method for solids is that the sample should exhibit photoacoustic signal saturation over the excitation wavelength range of interest.

References

[1] Harshbarger, W. R. and Robin, M. B., *Accounts of Chemical Research*, Vol. 6, 1973, p. 329.
[2] Rosencwaig, A., *Analytical Chemistry*, Vol. 47, 1975, p. 592A.
[3] Adams, M. J., King, A. A., and Kirkbright, G. F., *Analyst (London)*, Vol. 101, 1976, p. 73.
[4] Adams, M. J., Beadle, B. C., King, A. A., and Kirkbright, G. F., *Analyst (London)*, Vol. 101, 1976, p. 553.
[5] Callis, J. B., *Journal of Research of the National Bureau of Standards, Section A*, Vol. 80, No. 3, 1976, p. 413.
[6] Lahmann, W. and Ludewig, H. J., *Chemical Physics Letters*, Vol. 45, 1977, p. 177.
[7] Melhuish, W. H., *Journal of Physical Chemistry*, Vol. 65, 1961, p. 229.
[8] Tregellas-Williams, J., *Journal of the Electrochemical Society*, Vol. 105, 1958, p. 173.
[9] Lipsett, F. R., *Progress in Dielectrics (London)*, Vol. 7, 1967, p. 217.
[10] Wrighton, M. S., Ginley, D. S., and Morse, D. L., *Journal of Physical Chemistry*, Vol. 78, 1974, p. 2229.
[11] Rosencwaig, A. and Gersho, A., *Journal of Applied Physics*, Vol. 47, 1976, p. 64.
[12] McClelland, J. F. And Kniseley, R. N., *Applied Physical Letters*, Vol. 28, 1976, p. 467.
[13] Adams, M. J., Highfield, J. G., and Kirkbright, G. F., *Analyst*, Vol. 106, 1981, p. 850.

W. Rudolf Seitz,[1] Linda A. Saari,[2] Zhang Zhujun,[3] Steven Pokornicki,[4] Robert D. Hudson,[5] Steven C. Sieber,[6] and Mauri A. Ditzler[7]

Metal Ion Sensors Based on Immobilized Fluorogenic Ligands

REFERENCE: Seitz, W. R., Saari, L. A., Zhujun, Z., Pokornicki, S., Hudson, R. D., Sieber, S. C., and Ditzler, M. A., "**Metal Ion Sensors Based on Immobilized Fluorogenic Ligands,**" *Advances in Luminescence Spectroscopy, ASTM STP 863*, L. J. Cline Love and D. Eastwood, Eds., American Society for Testing and Materials, Philadelphia, 1985, pp. 63–77.

ABSTRACT: A metal ion sensor may be prepared by immobilizing a fluorogenic ligand on the common end of a bifurcated fiber optic. Fluorescence is excited through one arm of the fiber optic and observed through the other. When the immobilized ligand is placed in a solution of metal ion, some of the metal associates with the ligand causing its fluorescence characteristics to change. Equations are derived for the response of a sensor for systems where a 1:1 complex is formed.

We have successfully immobilized several ligands including morin, quercetin, and calcein to cellulose using cyanuric chloride as a coupling reagent. Immobilized morin is only very weakly fluorescent but forms fluorescent complexes with Al^{+3} and Be^{+2} and can be used to sense these ions. Immobilized calcein is fluorescent by itself but is quenched by several metal ions. We have also immobilized hydroxy napthol blue and 8-hydroxyquinoline-5-sulfonate on anion exchangers. In addition 2, 2', 4-trihydroxy-azobenzene and *p*-tosyl-8-aminoquinoline have been immobilized to silica gel.

KEY WORDS: detectors, optical detection, fluorescence, metal sensor, optical sensor, morin, calcein, cyanuric chloride, immobilization, fluorescence sensor, fluorogenic sensor, fluorogenic ligand, fluorescent complexes.

A metal ion sensor may be prepared by immobilizing a fluorogenic ligand on the common end of a bifurcated fiber optic. This type of sensor is illustrated

---

[1] Professor of chemistry, Department of Chemistry, University of New Hampshire, Durham, NH 03824.
[2] Research chemist, Instrumentation Laboratory, Inc., Lexington, MA 02173.
[3] Professor of chemistry, Shaansi Normal University, Sian, Shaansi, People's Republic of China.
[4] Chemist, C. R. Bard Corp., North Reading, MA 01864.
[5] Chemist, Texaco, Inc., Beacon, NY 12508.
[6] Student, Yale University Medical School, New Haven, CT 06511.
[7] Assistant professor, Department of Chemistry, College of the Holy Cross, Worcester, MA 01610.

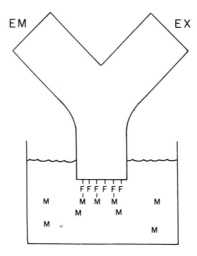

FIG. 1—*Diagram of metal ion sensor based on fluorescence. F is the immobilized fluorogenic ligand, M is the metal, and MF is the immobilized complex. EM = emission and EX = excitation.*

schematically in Fig. 1. Ligand F is immobilized on the common end of the bifurcated fiber optic. When the end of the fiber optic is immersed in a solution of metal ion M the metal associates with ligand establishing an equilibrium

$$M + F \rightleftarrows MF \tag{1}$$

If the fluorescence characteristics of MF differ from those of F, the relative amounts of MF and F can be determined by exciting fluorescence through one arm of the fiber optic and measuring fluorescence intensity transmitted through the other arm. This device is similar to a potentiometric electrode in that it responds to metal ion activity rather than total concentration and can be used for continuous sensing.

Many fluorogenic ligands for determining metal ions have been described [1–4]. In principle, any of these reagents may be immobilized to form a sensor. The two-phase configuration of the sensor enhances the utility of the reagent in several ways. Because the reagent is immobilized on the end of the fiber bundle it effectively preconcentrates the metal ion in the optical path, thereby enhancing sensitivity. Also, the sensor configuration eliminates or reduces optical interferences such as pre-filter and post-filter effects associated with sample absorbance or background signal caused by sample fluorescence. If conditions are properly chosen, the fluorescence sensor will extract only a very small fraction of the metal ion in the sample and thus will serve as a true sensor that does not perturb the composition of the sample. Furthermore, the sensor can be used for continuous measurements. Another advantage of the sensor is derived from its use of a fiber optic, which allows long-range transmission of the fluorescent signal in optical form. The possibilities of remote spectrometry based on fiber optics have attracted considerable attention recently in science news articles [5,6].

Fluorescence sensors also have some inherent disadvantages. Selectivity is limited by the selectivity of the ligand for the metal ion of interest. The sensor measurement will have a slower response time than a single-phase measurement because of the time required for mass transfer from the sample to the immobilized reagent phase. Many sensors will have a response that is inherently pH dependent because the metal ion displaces one or more protons when it forms the complex. Finally, to be practical for a sensor a fluorogenic ligand must be stable for extended time.

The development of sensors based on fluorescence is a new area of analytical chemistry. We have described a pH sensor based on immobilized fluoresceinamine [7] and an Al^{+3} sensor based on immobilized morin [8]. Other ongoing research in this area has been described in news articles [5,6]. Immobilized fluorogenic reagents have also been used for extraction and analysis in a nonsensor configuration [9,10]. This type of application is appropriate for immobilized ligands that bind metal ions too strongly to act as sensors.

This article will consider the theory of fluorescence sensors in some detail. It will briefly consider sensor design and instrumentation and will review progress to date in developing immobilized reagent systems.

Theory

The extent of metal association with immobilized ligand depends on the equilibrium constant for binding. If a one to one complex is formed by the immobilized ligand, then the equilibrium constant for binding may be represented

$$K_e = \frac{X_{MF}}{X_F[M]} \quad (1)$$

where X_F is the number of moles of immobilized ligand not associated with metal, X_{MF} is the number of moles of immobilized ligand associated with metal, [M] is the concentration of metal in solution (neglecting activity considerations), and K_e is the equilibrium constant for binding.

Since the total number of immobilized ligands is fixed

$$C_F = X_F + X_{MF} \quad (2)$$

where C_F is the total number of immobilized ligands in moles.

The fluorescence signal will depend on the relative amounts of F and MF

$$I_F = K_F X_F + K_{MF} X_{MF} \quad (3)$$

where I_F is the fluorescence intensity, and K_F and K_{MF} are proportionality constants relating fluorescence intensity to the amounts of free and associated ligand, respectively, assuming that the reagent layer does not absorb source radiation to a significant extent. If there is significant absorption in the reagent phase then an additional term is required

$$I_F = (K_F X_F + K_{MF} X_{MF})(1 - 10^{-A})/2.303A \quad (4)$$

where A is the absorbance of the reagent phase [*11*]. Both free and associated ligand can contribute to the overall absorbance

$$A = a_F X_F + a_{MF} X_{MF} \quad (5)$$

where a_F and a_{MF} are absorptivities for free and associated ligand, respectively. If a_F and a_{MF} differ, then the absorbance term will vary with the relative amounts of free and associated ligand and will influence the variation in fluorescence intensity with concentration of metal ion.

By substituting Eq 2 into Eq 1 and rearranging

$$X_F = \frac{1}{1 + [M]K_e} C_F \quad (6)$$

$$X_{MF} = \frac{[M]K_e}{1 + [M]K_e} C_F \quad (7)$$

Insertion of the sensor reduces the initial metal ion concentration $[M]_i$ because some of the metal is extracted into the reagent phase to form the immobilized complex

$$[M] = [M]_i - (X_{MF}/V) \quad (8)$$

where V is the volume of sample. It is possible to substitute for $[M]$ in Eqs 4 and 5 using Eq 8. This can be arranged and solved to yield an expressison for X_{MF} and X_F. These in turn can be inserted into Eq 3 to yield an expression for I_F as a function of initial metal concentration, which accounts for the depletion for metal ion in solution when the sensor is inserted in solution. This expression has several terms including a dependence on both the number of immobilized ligands C_F and the volume of sample. A considerably simpler expression is arrived at if it is assumed that analyte depletion in the sample is negligible

$$I_F = K_F \frac{1}{1 + [M]K_e} C_F + K_{MF} \frac{[M]K_e}{1 + [M]K_e} C_F \quad (9)$$

If only the complex fluoresces and ligand does not, then the first term drops out. In this case fluorescence is proportional to low metal concentrations. At higher concentrations, response becomes nonlinear ultimately reaching a limiting value when all immobilized ligands have combined with metal. This is shown in Fig. 2, which shows relative fluorescence as a function of metal concentration in terms of K_e. When the intensity is half the limiting value, $K_e = 1/[M]$. This provides a convenient means of estimating the value for K_e.

For the case where only the complex fluoresces and the first term of Eq 9 is zero, the remaining terms can be rearranged as follows

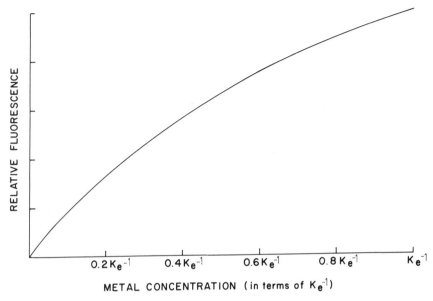

FIG. 2—*Theoretical variation in fluorescence intensity with metal concentration for a system in which the complex fluoresces and the ligand does not. Metal ion concentrations are expressed in terms of the equilibrium constant for binding* K_e.

$$\frac{[M]}{I_F} = \frac{[M]}{K_{MF}C_F} + \frac{1}{K_{MF}C_F K_e} \qquad (10)$$

A plot of $[M]/I_F$ versus $[M]$ will be linear with a slope of $1/K_{MF}C_F$ and intercept of $1/K_{MF}C_F K_e$. This type of plot may be used to verify that the above model applies in addition to providing a method for evaluating K_e.

In the region of linear response to metal ion concentration ($[M]K_e \ll 1$ and $X_{MF} \ll X_F$) the fraction of total metal ion extracted by the reagent phase is readily calculated

$$\frac{\text{Amount of metal in sample}}{\text{Amount of metal in reagent phase}} = \frac{X_{MF}}{[M]V} = \frac{K_e C_F}{V} \qquad (11)$$

This expression is arrived at by substituting for $X_{MF}/[M]$ using Eq 1 and assuming $C_F = X_F$. This expression provides a convenient means of determining whether or not a particular system is acting as a sensor and not extracting a significant amount of metal from the sample. It also shows how the fraction of metal extracted depends both on the number of immobilized ligands and the volume of solution.

If the ligand fluoresces by itself, but not in a complex then the second term of Eq 9 drops out. In this case, fluorescence intensity decreases with added metal ion. Figure 3 shows the expected response with metal concentration expressed in terms of K_e. This kind of system is less desirable for analytical purposes

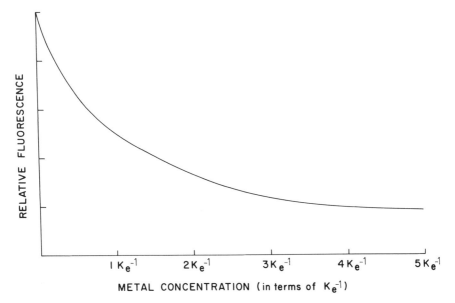

FIG. 3—*Theoretical variation in intensity with metal concentration for a system in which ligand fluorescences and complex does not. Metal concentrations are expressed in terms of the equilibrium constant for binding* K_e.

because it is less sensitive and has limited dynamic range. However, it is the only means of designing systems that respond to metal ions that tend to quench fluorescence.

Many fluorogenic reagents for metal ions form 2:1 complexes in solution. This could also happen in the immobilized reagent phase. This is more difficult to deal with theoretically. In writing Eq 1, for a 1:1 complex it is tacitly assumed that activity effects in the immobilized phase are the same for both free and associated ligand and thus cancel out. However, for a 2:1 complex these do not cancel because the expression for ligand must be squared.

A further difficulty in dealing with 2:1 complexes involving immobilized ligands is that steric factors will play a major role influencing the stability of the 2:1 complex. These steric factors could vary greatly location to location on the immobilized phase.

If the following expression is written for the formation of a 2:1 complex

$$K_e = \frac{X_{MF_2}}{X_F^2 \cdot [M^{+n}]} \quad (12)$$

then it is possible to derive an expression analogous to Eq 9 for fluorescence intensity as a function of methods in concentration. The expressions are more complex but yield results similar to the one to one case. If only the complex fluorescences, the response to metal ion is linear at low concentrations and ultimately reaches a saturating value. The shape of the response as saturation is approached will differ from the one to one case. It is important to keep in mind that K_e as defined in Eq 12 has no thermodynamic significance whatsoever and would be expected to vary with the amount of MF_2.

Design Considerations

There are several factors to be considered when immobilizing a fluorogenic ligand to serve as a sensor. Some of these are considered below.

1. Amount of Ligand: Eq 9 indicates that fluorescence intensity if proportional to the total number of immobilized ligands. In practice, however, concentration quenching may occur when ligands are spaced too closely together causing a reduction in intensity. This is observed with fluorescein derivatives, for example. Increasing the amount of immobilized ligand also causes a higher percentage of metal to be extracted from the solution and makes it harder to formulate the reagent phase to permit rapid mass transfer to the immobilized ligands. Accordingly, the amount of ligand immobilized should be the smallest possible amount that yields a sufficiently large fluorescence signal for the intended application. If more ligand is immobilized, the inner filter effects become more severe, affecting the shape of intensity versus metal concentration curves. Normally this would be undesirable, however, a highly absorbing reagent phase can serve as a means of screening the sensor from background fluorescence from the sample.

2. Immobilization Procedure: Immobilization necessarily introduces a new functional group to the ligand. This may affect the fluorogenic properties of the ligand. For example, 8-hydroxyquinoline immobilized to solid supports via an azo coupling does not form fluorescent complexes unlike the free ligand. The reason for this is that the azo group introduces a low energy n-π^* transition into the ligand. As will be illustrated in the next section, immobilization can also have the reverse effect causing an otherwise nonfluorescent ligand to become fluorescent.

In our work, we have had success immobilizing various ligands via cyanuric chloride [7,8,12]. This is a convenient, versatile reagent that couples both to amino and phenolic compounds. However, the linkage is not stable at high pH.

3. Substrate for Immobilization: To minimize the amount of reflected excitation radiation the substrate should present as few surfaces as possible to the excitation radiation and should have a refractive index similar to that of the sample. From an optical point of view, a clear film is the best type of substrate.

The substrate should also be designed to facilitate mass transfer of metal to the immobilized ligand. For mass transfer, a film is less satisfactory because inside the film only relatively slow diffusive mass transfer can take place. The film needs to be as thin as practical to get an adequate response time. A fine powder with ligand immobilized on the surface is more suitable for mass transfer than a film.

Most of our work has involved powdered cellulose as the substrate. Although this has been satisfactory in most respects, it does lead to a significant background because of reflected excitation radiation.

Instrumentation

Other than the fiber optic itself, the metal ion sensors do not require specialized instrumentation. Our instrumentation involves a tungsten-halogen source, a bifurcated fiber optic, filters for wavelength selection, and a photomultiplier. It has been described in more detail elsewhere [7]. The bifurcated fiber optic can be used in conjunction with standard spectrofluorometers by adapting the sample compartment so that excitation and emission ends of the fiber optic can be appropriately placed with respect to excitation and emission optics.

Because the bifurcated fiber optic has separate fibers, it requires that the immobilized reagent be a certain distance from the actual optic surface for fluorescence to be observed. The reason for this is illustrated in Fig. 4. Fluorescence is only observed for those immobilized ligands that are in both the cone of excitation radiation emerging from the excitation fibers and in the cone corresponding to the range of angles over which the emission fibers accept light. Fluorescence is not directly observed from immobilized ligands that are too close to the optic surface.

A single-fiber optic can be used as well [6]. In this case excitation and emission radiation travel along the same fiber optic. A beam splitter directs part of the emitted radiation to the detection optics, which resolve fluorescence from scattered excitation radiation on the basis of wavelength. The single-fiber optic approach is clearly preferred for remote sensing, which requires a long optic.

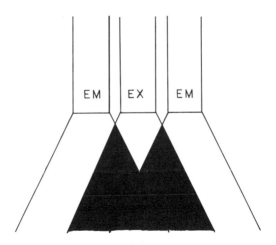

FIG. 4—*Closeup view of fiber optic showing one fiber carrying excitation radiation (EX) and two fibers carrying fluorescence emission (EM). For fluorescence to be directly observed, the fluorophase must be within both the cone on radiation emitted from the excitation fiber and the case of radiation accepted by the emission fibers. This area is shaded.*

Immobilized Ligands

We have successfully immobilized several ligands. Some of these have or promise to yield useful sensors. In other cases, difficulties have been encountered that preclude sensor development. Results with various ligands are summarized below:

Morin

Morin (3,4,7,2',4'-pentahydroxyflavone) has been immobilized to cellulose via cyanuric chloride. Since cyanuric chloride activated cellulose can be coupled to any of the five hydroxy groups, the product is actually a complex mixture. The batch of immobilized morin that was characterized in greatest depth was found to have 4.2×10^{-5} mol of morin per gram of cellulose. However, the same batch had a binding capacity of only 1.1×10^{-5} mol Al^{+3} per gram of cellulose, indicating that for much of the morin, the immobilization procedure ties up the hydroxy group involved in complexation.

Even though only about a quarter of the immobilized morin is able to form complexes, the immobilized morin can serve as the basis of a useful sensor. The immobilized ligand fluoresces only very weakly by itself and forms strongly fluorescent complexes with both Al^{+3} and Be^{+2}. The excitation spectrum for the immobilized Al^{+3}-complex is broader than the solution spectrum and shifted to slightly longer wavelengths. The excitation spectrum for immobilized Be^{+2} complex at pH 5 shows a new band not observed in solution at that pH. Response to both ions is linear at low concentrations and curves, ultimately reaching a saturation value at high metal concentration. Typical data are shown in Figs. 5 and 6. These data can be replotted as $[M]/I_F$ versus $[M]$ to determine whether Eq 10 applies. These plots come out quite linear confirming 1:1 stoichiometry. The binding constants are on the order of 1×10^4 for Al^{+3} and 9×10^3 for Be^{+2} at pH 5. They vary with pH in the manner expected from the solution chemistry of Al^{+3} and Be^{+2}. Under typical operating conditions (for 1 mg of immobilized ligand and 15 cm³ of metal ion solution only 1% or less of the metal is extracted into the reagent phase). In the case of Be^{+2} one would anticipate stronger binding and greater sensitivity in basic solutions. Unfortunately, under these conditions, hydrolysis causes the ligand to become detached from the substrate. This represents a limitation of immobilization based on cyanuric chloride.

The selectivity of the immobilized ligand is similar to that for the ligand in solution. Ferric iron is the most serious interference, forming strong complexes that do not fluoresce.

The response of immobilized morin to Al^{+3} has been described elsewhere [8]. We expect to report on the response to Be^{+2} in more detail in the near future.

Quercetin

Quercetin (3,5,7,3',4'-pentahydroxyflavone) has been immobilized by the same procedure as morin. Immobilized quercetin responds to Al^{+3} in much the

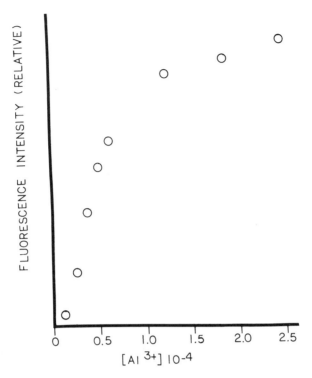

FIG. 5—*Fluorescence intensity versus aluminum concentration for sensor based on immobilized morin at pH 4.8.*

same manner as immobilized morin. The batch of immobilized quercetin prepared by us was less sensitive to Al^{+3} than morin. This may reflect a difference in the amount of ligand immobilized rather than an intrinsic difference in sensitivity. Nonetheless, because of its expected similarity to morin, immobilized quercetin has not been characterized in depth.

Hydroxynaphthol Blue

Hydroxynaphthol blue (HNB)(I) belongs to the 0,0'-dihydroxyazobenzene class of ligands, which are nonfluorescent by themselves, but form fluorescent complexes with several metals [3]. The three sulfonate groups allow for easy immobilization onto an anion exchanger. HNB was immobilized on an anion exchange membrane (R-1035 from RAI Research Corporation) with a thickness of 0.05 mm. The immobilization procedure is simply to immerse the membrane in a solution of HNB. The membrane completely extracts all HNB from solution, as long as the amount of HNB does not exceed the number of anion exchange sites. Thus the amount of HNB/cm² of membrane is readily controlled by appropriately adjusting membrane area and the concentration and volume of HNB solutions.

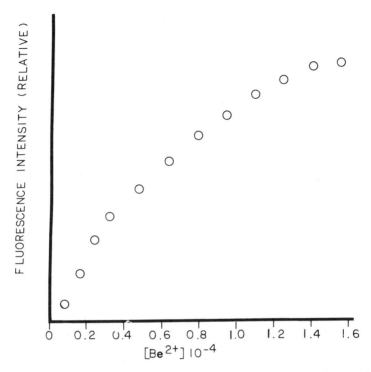

FIG. 6—*Fluorescence intensity versus beryllium concentration for sensor based on immobilized morin at pH 5.2.*

The response of immobilized HNB to added Al^{+3} is most interesting. When Al^{+3} is added to freshly immobilized HNB at pH 5, there is an essentially instantaneous change in color from blue to red. The red complex fluoresces red while the blue complex does not fluoresce. The complex fluorescence appears to be subject to concentration quenching since fluorescence intensity is greater for smaller amounts of HNB. Unfortunately, the rapid response to added Al^{+3} is only observed the first time the immobilized reagent is contacted with solution. Response rapidly falls off and becomes very slow kinetically. In fact, the membrane has to be soaked in ethylenediaminetetraacetate (EDTA) for extended periods of time to removal all the metal ion. The behavior indicates that electrostatic effects are influencing mass transfer in the membrane. The initially rapid response suggests that after immobilization of HNB there is an excess of fixed negative charges in the membrane. Since the membrane itself is cationic, the negative charges are presumably caused by sulfonate groups on HNB, which for steric reasons are not associated with cationic sites in the membrane. This creates an electric field that causes Al^{+3} to migrate into the membrane. However, once the complex forms, the excess anionic sulfonate sites are apparently neutralized leading to a situation where electric field effects tend to prevent Al^{+3} from entering the membrane. As a consequence, HNB immobilized in an ion exchange membrane is not suitable for fluorescence sensing although it may have value as a reagent used on a one-time disposable basis.

8-Hydroxyquinoline-5-Sulfonate

8-Hydroxyquinoline-5-sulfonate (8HQ5S), like HNB may be immobilized on anion exchangers. However, because there is only one sulfonate group versus three for HNB, 8HQ5S is more easily displaced from ion exchangers by other anions.

When 8HQ5S is immobilized on an anion exchange membrane, it behaves similarly to HNB. Initial response to added metal is rapid, but soon thereafter response becomes exceedingly slow.

8HQ5S can also be immobilized on an anion exchange resin. In this case, response is not affected by the kind of electrostatic effects observed with the membrane. 8HQ5S immobilized on an anion exchange resin is nonfluorescent by itself but forms fluorescent complexes with Cd^{+2}, Zn^{+2}, and other metal ions. This appears to be a useful system and is currently undergoing characterization.

Salicylidene-o-Aminophenol

2,4-Dihydroxy benzene-*o*-aminophenol was prepared by condensing 2,4-dihydroxybenzaldehyde with *o*-aminophenol. The ligand in solution was only weakly fluorescent but formed strongly fluorescent complexes with zinc, magnesium, and aluminum. Maximum emission for the complex occurred at 488 nm.

When 2,4-dihydroxybenzene-*o*-aminophenol was immobilized to cellulose using cyanuric chloride, the product was found to fluoresce bright orange. We are at a loss to explain this unexpected observation. This product conceivably could be used to sense metal ions based on quenching. However, this particular possibility has not been pursued further.

Calcein

We have immobilized calcein to cellulose using the cyanuric chloride procedure. Calcein in solution may be used as a reagent for calcium determination because at high pH it is nonfluorescent by itself but forms fluorescent complexes with calcium. This is not possible with calcein immobilized via cyanuric chloride because calcein rapidly hydrolyzes away from the substrate at pHs of the calcium determination. However, at neutral pHs the immobilized calcein may be used to detect a variety of metals based on fluorescence quenching. We have characterized response to Ni^{+2}, Cu^{+2}, and Fe^{+3} all of which quench calcein fluorescence. The fluorescence intensity decreases with added metal as illustrated in Fig. 7. Binding constants were measured by a competitive binding procedure. Iminodiacetate is added to a solution of metal ion quencher in contact with immobilized calcein. When the added iminodiacetate concentration is large enough, it pulls the metal quencher away from the calcein causing the calcein to become fluores-

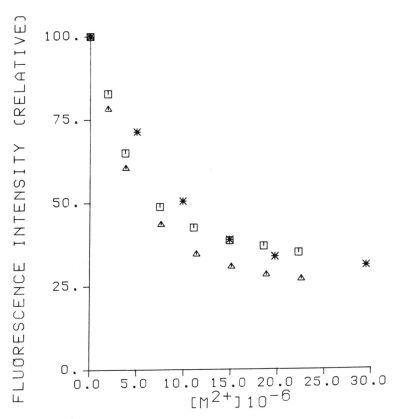

FIG. 7—*Fluorescence intensity versus metal concentrations for immobilized calcein at pH 7.0.* $\Delta = Cu^{+2}$ □ $= Co^{+2}$ * $= Ni^{+2}$.

cent. The conditional equilibrium constant for formation of calcein-metal complexes may be calculated from known constants for iminodiacetate-metal complexes and the experimental value of the concentration required to pull metal from calcein.

The binding constants for immobilized calcein complexes are quite large. At pH 7, the values for Ni^{+2}, Cu^{+2}, and Co^{+2} are 1×10^{11}, 3×10^{12}, 1×10^{8}, respectively. Because the constants are so large, immobilized calcein does not act as a sensor. Instead, it will extract a large percentage of metal ion from a solution unless the metal is strongly complexed.

Although immobilized calcein is not suitable for a sensor, it can be used for other purposes. It can be used to detect endpoints of compleximetric titrations. It can also be used as a reagent to preconcentrate metal ions from solution for subsequent analysis. The fluorescence characteristics of calcein provide a convenient means of determining whether or not a given batch of immobilized calcein is saturated with metal. Finally, immobilized calcein can be used for analysis based on quenching, provided the sample contains only one metal. We plan a more complete report on immobilized calcein in the near future.

2,2',4-Trihydroxyazobenzene

This compound, like several others containing the o,o'-dihydroxyazo functional group, forms fluorescent complexes with Al^{+3}. The additional, noncomplexing hydroxy group was used to couple the ligand to silica gel through a cyanuric chloride linkage. The immobilized ligand alone is nonfluorescent but the Al^{+3} complex exhibits fluorescence that closely mimics the nonbound complex. Like calcein, this ligand binds the metal ion so tightly that it is not well suited for use on a sensor. We have however found it to be useful for preconcentrating the metal before detection. In this application the substrate is not attached to a fiber optic; rather, it is added directly to the solution then, after equilibration, it is isolated by filtration. Fluorescence is measured from the front surface of a packed powder cell.

The principal analytical advantage of this approach is its potential to decrease interference problems. After isolation from the solution matrix the substrate-complex system can be washed free of potentially interfering species. The tightly held metal ion is not lost during this process [9].

Preliminary work suggests that this immobilized complex may be useful for the determination of fluoride. There exist a number of procedures for the analysis of F^- based on its removal of Al^{+3} from fluorogenic ligands and their subsequent decrease in fluorescence [1]. We have observed a similar phenomenon for the immobilized ligand-Al^{+3} system. That is, equilibration with solutions containing traces of F^- diminish its subsequent fluorescence. The possibility of using this immobilized ligand for indirect determination of F^- is currently under investigation.

p-Tosyl-8-Aminoquinoline

This ligand has been reported to form fluorescent complexes with Zn^{+2} and Cd^{+2}. In order to immobilize a system containing the chelating portions of this ligand we synthesized an analog with n-propionic acid in place of the methyl group. The acid was then refluxed with n-propyl amine immobilized on silica gel. In the presence of the dehydrating agent, N,N'-dicyclohexylcarbodiimide (DCC), the corresponding amide formed, thereby linking the ligand to the silica gel.

The immobilized ligand forms an intensely fluorescent complex with Zn^{+2} [10]. At this time the complex has been examined only by front surface fluorescence as the isolated powder, that is, it has not yet been attached to a sensor. In this form the fluorescence intensity is linearly related to Zn^{+2} concentration virtually to the point of saturation. This comes at 2.5 mg/dm^3 when 0.25 g of substrate is equilibrated with 0.25 dm^3 of solution. Zn^{+2} concentrations as low as 0.03-ppm Zn^{+2} can be detected with these quantities. The modest fluorescence of the immobilized ligand alone is a disadvantage. The ability of several transition metals to decrease the fluorescence, probably by displacing the Zn^{+2}, is also a disadvantage. Further work on the optimization and characterization of this system is in progress and will be reported at a later time.

Acknowledgment

Partial support for this research was provided by National Science Foundation (NSF) Grant CHE-820613.

References

[1] White, C. E. and Argauer, R., *Fluorescence Analysis, a Practical Approach*, Marcel Dekker, New York, 1970.
[2] Wehry, E. L., *Analytical Chemistry*, Vol. 56, No. 5, April 1984, pp. 156R–173R and Vol. 54C, No. 5, April 1982, pp. 131R–150R.
[3] Seitz, W. R., *C.R.C. Critical Review of Analytical Chemistry*, Vol. 8, No. 4, April 1980, pp. 387–406.
[4] Guilbault, G. G., *Practical Fluorescence Theory, Methods and Techniques*, Marcel Dekker, New York, 1973, Chapter 6.
[5] Maugh, T. H., II, *Science*, Vol. 218, No. 4575, Nov. 1982, pp. 875–876.
[6] Borman, S. A., *Analytical Chemistry*, Vol. 53, No. 14, Dec. 1981, p. 1616A.
[7] Saari, L. A. and Seitz, W. R., *Analytical Chemistry*, Vol. 54, No. 4, April 1982, pp. 821–823.
[8] Saari, L. A. and Seitz, W. R., *Analytical Chemistry*, Vol. 55, No. 4, April 1983, pp. 667–670.
[9] Ditzler, M. A., Doherty, G., Sieber, S., and Allston, R., *Analytica Chimica Acta*, Vol. 142, Oct. 1982, pp. 305–311.
[10] Ditzler, M. A., *Trends in Analytical Chemistry*, Vol. 2, No. 5, May 1983, pp. viii–ix.
[11] Holland, J. F., Teets, R. E., Kelly, P. M., and Timnick, A., *Analytical Chemistry*, Vol. 49, No. 6, May 1977, p. 706.
[12] Kay, G. and Crook, E., *Nature*, Vol. 216, Nov. 1967, pp. 514–515.

Manipulation of Luminescence Observables

Francis J. Purcell,[1] Raymond Kaminski,[1] and
Ralph H. Obenauf[1]

Synchronous-Excitation Fluorescence Applied to Characterization of Phenolic Species

REFERENCE: Purcell, F. J., Kaminski, R., and Obenauf, R. H., "**Synchronous-Excitation Fluorescence Applied to Characterization of Phenolic Species,**" *Advances in Luminescence Spectroscopy, ASTM STP 863,* L. J. Cline Love and D. Eastwood, Eds., American Society for Testing and Materials, Philadelphia, 1985, pp. 81–94.

ABSTRACT: Synchronous-excitation fluorescence (SEF) spectroscopy promises to be a quick, sensitive, and highly selective technique for the determination of hazardous contaminants. This work concentrates on detection of the phenols that pervade many areas of everyday life. Phenols frequently found in streams and groundwater as industrial pollutants also result as degradation products of green plants. The primary concern for these phenols is their known carcinogenic nature. In conjunction with derivative spectroscopy, synchronous-excitation fluorescence detects phenols in the parts per billion range and discriminates between common variants. By scanning the excitation and emission monochromotors with a 3-nm offset between them, each phenol species is characterized by a single peak. The water-Raman band, the prime interference at dilute concentrations in normal fluorescence, is also eliminated with the synchronous technique. Further, when derivatization is applied to the data, isomers, whose synchronous peaks occur within 5 nm of one another, can be separated from a mixture.

KEY WORDS: fluorescence, phenols, spectroscopy, synchronous-excitation fluorescence, derivative spectroscopy, limits of detection

Phenols seem to be everywhere (Fig. 1). The nuts and cereals we eat, the coffee, wine, and beer that we drink, all contain significant amounts of these aromatic alcohols. Synthetic phenols act as binders in plywood and fiberglass insulation, and they are employed in the manufacture of pharmaceuticals, dyes, and pesticides. Phenols are also by-products of coke production, paper processing, and coal liquefaction, and as a result of industrial pollution and the biodegradation of plants and algae, often surface in our natural water supplies.

---

[1]Manager of applications laboratory, spectrometers specialist, and manager of laboratory equipment division, respectively, SPEX Industries, Inc., 3880 Park Ave., Edison, NJ 08820.

Phenol

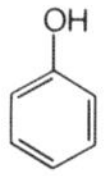

OCCURENCE

Food stuffs such as nuts, cereals, rice. Also coffee, wine, beer

Natural water supplies and hormones of animals and plants

Synthetic binders for plywood and insulation

HAZARDS

Phenol, parachlorophenol and others shown to encourage carcinogenic and teratogenic activity

Odors and tastes make drinking water unpalatable. WHO recommends > 1 ppb phenols in drinking water

BENEFITS

Ellagic acid neutralizes carcinogenic activity

Pharmaceuticals such as acetaminophen

FIG. 1 — *Phenols: structure, occurrence, hazards, and benefits.*

Furthermore, phenolic hormones play pivotal roles in our metabolism, nutrition, and even sexuality. Because of this frequent contact with such a ubiquitous class of compounds, their potential for harm or benefit raises some critical questions. Unfortunately, the answers are not clear cut.

Upon chlorination of drinking water supplies, trace amounts of phenols form odiferous and foul-tasting chlorophenols that may render the water unpalatable. Their well-established carcinogenic nature [*1–3*] prompted the World Health Organization (WHO) to limit the total phenolic content of drinking water to 1 ppb [*3*]. This agrees with the Environmental Protection Agency (EPA) specifications for domestic drinking water supplies as well as for protection against tainting of fish flesh [*4*].

On the other hand, many phenolic compounds are benevolent. Acetaminophen, a routine non-aspirin painkiller, is a phenol. Another, ellagic acid, has demonstrated such a decisive ability to neutralize the active forms of the carcinogen benzo(*a*)pyrene that it is considered to be the prototype for a new class of cancer-inhibiting drugs [*5,6*].

This mixture of hazards and virtues related to phenolic compounds emphasizes that it is not sufficient to determine the total phenolic content of a sample. It is especially crucial to establish which species are present.

Quite predictably, chromatographic techniques [7-9] have been favored for their ability to separate not only phenols from one another, but also to weed out the inevitable, interfering components of real-world samples. Although ultraviolet (UV) absorption and refractive index detectors have been widely applied to liquid chromatography (LC), fluorescence because of its sensitivity, selectivity, and convenience, has become a common detection technique for phenols. One study [9] of phenol and alkyl phenols showed that fluorescence LC detectors were two to twelve times more sensitive than conventional UV detectors. The detection limits in that study were about 3 ppb.

Other researchers have applied synchronous scanning and second derivatives [10] to phenol mixtures in an effort to separate out and identify the contributions of each component while completely bypassing LC.

In this presentation, we take fluorescence one step further, demonstrating that it is feasible to identify and quantify phenols with fluorescence alone, since now synchronous scanning, high-order derivatives, and other specialized data-processing routines have become a standard part of commercial spectrofluorescence systems. The real bonus of this alternative is realized in sample handling; the most time-consuming procedure is filling the cuvette with the substance to be analyzed.

Instrumentation

The spectrofluorometer used in these studies is diagrammed in Fig. 2. The SPEX Fluorolog® 2 is a modular system, in this case incorporating single grating spectrometers and a single-beam sample module. Control of the system is through the programmable Datamate multiprocessor, which combines the photometers, high voltage (HV) supplies, scan control, and data-processing facilities in a single unit.

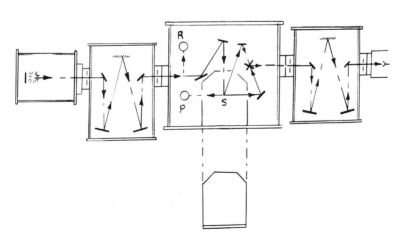

FIG. 2—*Optical diagram of F111 Fluorolog.*

The spectra for this study were scanned on a F111 Fluorolog consisting of single spectrometers for excitation and emission wavelengths and a single-beam sample module. Photon-counting detection was used throughout, the signal being integrated for 1.0 s at each point in a scan. The detection limits quoted could, obviously, be extended simply by setting longer integration times.

Fluorescence Detection

Fluorescence has two inherent advantages over absorption techniques. The first is discrimination. Although most compounds absorb in the UV, complicating their identification in multicomponent mixtures, not all can be expected to fluoresce. The second advantage is in detectivity. Noise is the limiting factor. In fluorescence, the detected signal is essentially proportional to sample concentration, and thus impurity background limited. On the other hand, an absorption measurement takes the difference between two large signals and therefore is detector-noise limited. As a consequence, the associated noise is much higher.

The prime disadvantage of fluorescence is that the peaks are usually broad, so any complex mixture can expect to suffer from significant overlap of spectral features. By exciting a real-world sample with constant wavelength radiation, the observed spectrum would probably look something like the one shown in Fig. 3. The solid line is one scan of a mixture of some typical polynuclear aromatic (PNA) compounds. If instead, the excitation and emission spectrometers are scanned together with a constant wavelength offset between them, the components of the mixture may drop out into unique sharp peaks, much like those

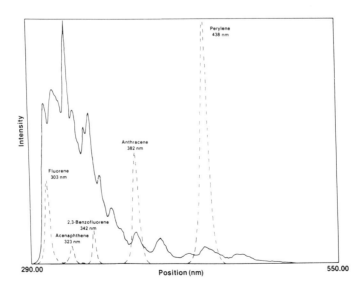

FIG. 3 — *Synchronous scanning resolved the spectrum of a mixture of poly-nuclear aromatics into five distinct components* (dashed line) *though a normal emission scan* (solid line) *could not.*

of a chromatogram. This is the synchronous-excitation fluorescence technique (SEF).

Synchronous Scanning

Originally applied to fingerprint complex samples, such as crude oils [11,12], synchronous scanning has more recently gained wider acceptance [10,12,13]. Briefly, synchronous scanning gives a view of the overlap of emission I_m and excitation spectra I_x. The recorded intensity I_s is actually proportional to the product of both [12]

$$I_S \propto I_X(\lambda) \times I_M(\lambda + \Delta\lambda)$$

where λ is wavelength in nm. A comparison between excitation, emission, and synchronous spectra of phenol is shown in Figure 4 [10].

The new parameter $\Delta\lambda$ introduces an added variable into our detection equation. In most cases, the greatest detectivity is obtained with $\Delta\lambda$ about equal to the Stokes shift, or $\Delta\lambda = (\lambda M_{max} \lambda X_{max})$; while the greatest selectivity is obtained with $\Delta\lambda$ just greater than the bandpass of excitation and emission spectrometers, that is, large enough to keep the Rayleigh scatter from passing the excitation radiation through the emission spectrometer.

The effectiveness of this selection process on a mixture of phenols is illustrated in Fig. 5. A mixture of 10-ppm phenol, 8-ppm p-cresol, 2-ppm 2-naphthol, and 3.5-ppm 4,5-dihydroxynaphthalene in aqueous solution emitted the spectrum shown as a dashed line when excited with fixed excitation wavelength. By synchronous scanning with a constant 3-nm offset between excitation and emission spectrometers, all four components could be readily identified by their characteristic peaks.

Besides separating components by their individual emissions, synchronous scanning also eliminates interference from solvents that may be especially critical

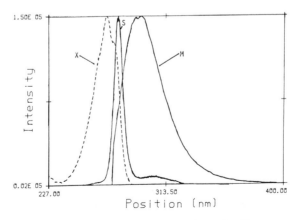

FIG. 4—*Comparison of fluorescence excitation* X *and emission* M *spectra to a synchronous-excitation spectrum* S *where the intensity* I_s *varies as* $I_s = I_x(\lambda)*I_m(\lambda + \Delta\lambda)$ *where* $\Delta\lambda = \lambda_m - \lambda_x$.

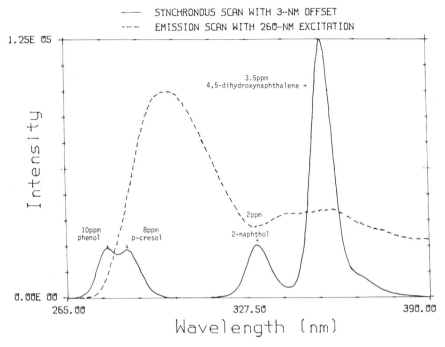

FIG. 5—*An emission scan of a sample at 260 nm (dashed line) fails to separate the four components of a mixture, though a synchronous scan with a 3-nm offset clearly identifies each phenol.*

at low sample concentrations, hence detectivity can be orders of magnitude better.

How do we determine the best offset for a particular sample? Three-dimensional scanning is one alternative. Figure 6 is a three-dimensional plot of 20-ppb phenol recorded and output by the Datamate multiprocessor. The diagonal axis traces the evolution of a synchronous scan as $\Delta\lambda$ is varied. Notice that increasing the offset between excitation and emission spectrometers allows the water-Raman scatter to overwhelm the sample signal at about $\Delta\lambda = 20$ nm. From that point on, the water-Raman scatter, which occurs at a constant energy difference from the excitation radiation rather than a constant wavelength, begins to move toward higher wavelengths until the two peaks separate once again. Juggling the $\Delta\lambda$ parameter in this way adds another dimension to fluorescence selectivity.

Table 1 lists the estimated limits of detection determined for 25 different phenolic species with an integration time of 1 s. The limit of detection is assumed to be a signal to noise ratio of 3 to 1. Each was computed with ordinary phenol as the standard. Actual spectra were obtained on phenol concentrations down to 1 ppb in water. The rest of the species were scanned at higher concentrations and

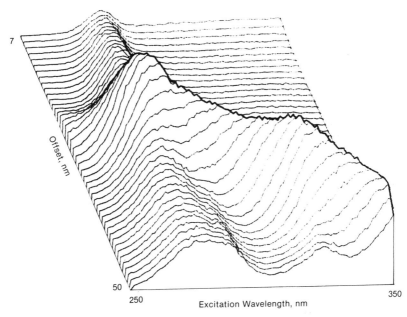

FIG. 6—*This three-dimensional synchronous scan of phenol (20 ppb in H_2O) demonstrates how the water-Raman scatter gradually moves toward higher wavelength as the offset increases. The interference is thereby eliminated.*

the limits of detection extrapolated. A linear response was assumed and was verified for phenol between 10 ppm and 1 ppb.

The estimated limits of detection given in Table 2 are those that can be expected from a single, rapid survey scan at a fixed offset of 3 nm again with a 1-s integration time. Of course, after screening a sample in this manner, individual peaks will have been identified. The offset can then be adjusted under program control to optimize parameters until the detection limits of Table 1 are achieved. In general, selecting $\Delta\lambda$ equal to the Stokes shift of one component will give best separation of mixtures, though this rule must be tempered by possible interference from sources such as water-Raman.

Derivative Spectroscopy

Though SEF presents a potent method for isolating fluorescing components in complex samples, additional discrimination is often necessary when only a few nanometres or so separates the positions of synchronous peaks. In such cases, derivatives may be applied to the results for enhanced resolution [10,14,15]. Two factors dominate the choice of derivative order. In general, enhancements of

TABLE 1 — *Estimated limits of detection for 25 phenols with an offset equal to the difference between excitation and emission peaks. Synchronous scans in each case.*

Compound	ELD, ppb[a]	Offset, nm
Phenol	1	22
2-Naphthol	0.3	27
Resorcinol	1	25
p-Cresol	10	30
o-Cresol	1	23
2,3-Dimethylphenol	0.2	24
m-Cresol	3	21
2,5-Dimethylphenol	1	27
2,6-Dimethylphenol	3	22
3,4-Dimethylphenol	1	28
3,5-Dimethylphenol	4	20
2,3,5-Trimethylphenol	10	28
2,3,6-Trimethylphenol	10	29
2,4,6-Trimethylphenol	5	28
o-Ethylphenol	2	26
m-Chlorophenol	150	27
p-Ethylphenol	10	27
p-Chlorophenol	10	30
2,3,4-Trichlorophenol	200	67
p-Bromophenol	500	42
o-Bromophenol	150	27
m-Nitrophenol	100	48
3,4-Dinitrophenol	50	57
Catechol	3	31
4,5-Dihydroxynaphthalene	2	133

[a] ELD is estimated limits of detection.

resolution increase with derivative order. Also, the fourth derivative has the property of associating a prominent maximum with the location of a spectral peak. This means that any existing peak-finding routine can be applied to fourth derivative spectra without modification.

In the Datamate data system, derivatives are keystroke-activated, variable-point Savitsky-Golay routines that are applied off line [16]. To form the fourth derivative, the second derivative is applied twice.

Phenol, o-cresol, m-cresol, and p-cresol have synchronous peaks that occur within 6 nm of one another, and though Fluorolog's optical resolution is more than capable of separating features that close, the small slits required would severely curtail detection limits. With large slits, on the other hand, none of the components of a mixture of these phenols is resolved by synchronous scanning alone (Fig. 7). Yet, the fourth derivative unequivocally locates all four.

Automation

To consolidate the full potency of the techniques of synchronous scanning and derivative analysis, a program such as the one shown in the flow chart of Fig. 8

TABLE 2—*Estimated limits of detection for 25 phenols with 3-nm offset. Synchronous scans with $\Delta\lambda = 3nm$ in each case.*

Compound	ELD, ppb[a]	Position, nm
Phenol	5	279
2-Naphthol	1	331.5
Resorcinol	5	281.5
p-Cresol	5	286.5
o-Cresol	5	280.5
2,3-Dimethylphenol	1	280
m-Cresol	10	282
2,5-Dimethylphenol	5	282.5
2,6-Dimethylphenol	10	278
3,4-Dimethylphenol	5	287
3,5-Dimethylphenol	10	281
2,3,5-Trimethylphenol	20	281
2,3,6-Trimethylphenol	40	280
2,4,6-Trimethylphenol	20	286
o-Ethylphenol	10	281
p-Ethylphenol	3	286
m-Ethylphenol	3	280
m-Chlorophenol	100	282.5
p-Chlorophenol	15	290.5
2,3,4-Trichlorophenol	30	334
p-Bromophenol	200	370
o-Bromophenol	1000	279
m-Nitrophenol	300	336
3,4-Dinitrophenol	200	331.5
Catechol	20	286
4,5-Dihydroxynaphthalene	5	354

[a]ELD is estimated limits of detection.

can be assembled so a single RUN command on the processor's keyboard is enough to identify and quantify the phenols present in a given sample.

The first step is to synchronously scan the desired region, then calculate the fourth derivative. Next, the lowest wavelength peak is located, and should its wavelength position match that of a known phenol, a previously recorded standard spectrum is read in from disk storage. The sample spectrum is then further examined for additional peaks, and standard spectra read from a disk for each component identified.

With this information, curve fitting begins. The operator may specify a numerical criterion for goodness of fit, and once that point is reached the actual concentration of each phenol species is derived from the calculated contribution of the standard spectrum to the result. The concentrations are then displayed on the cathode-ray tube (CRT) or output to a printer or other peripheral.

The performance of the program can be assessed from this synchronous scan of a mixture of phenols shown in Fig. 9. The fitted curve (dashed line) matches the raw data to within the operator-selected 10%. The individual standard spectra of the components, scaled to their proper relative contributions, are also given. Concentrations of 1.4-ppm phenol, 0.4-ppm *o*-cresol, 1.6-ppm *m*-cresol, and

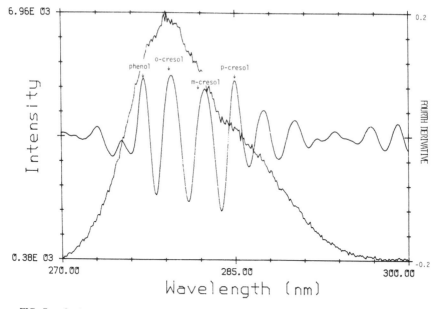

FIG. 7—*On its own, even a synchronous scan is not powerful enough to separate a mixture of phenol, o-cresol, m-cresol, and p-cresol into its components. Yet the application of the fourth derivative distinctly defines each contribution.*

2-ppm *p*-cresol were determined by the program. The mixture was crudely prepared to be 1 ppm in each component, thus the calculated fit is quite reasonable.

Real-World Samples

Of course, the utility of any analytical technique is severely curtailed if it requires that all samples be in distilled water or neat solvents, thus, several real-world samples were investigated, which included a variety of potential interferents. Figure 10 shows an emission and a synchronous scan of a water sample taken from the Raritan Bay just off the south shore of Staten Island. Since no native phenols were detected, the sample has been spiked with 200-ppb phenol, 200-ppb 2-naphthol, and 20-ppb 4,5-dihydroxynaphthalene. As might be expected, the emission scan is unintelligible while the synchronous excitation scan clearly separates the three components.

A further improvement of the results can be obtained by removing the background emission of the sample. This is conveniently accomplished by subtracting the spectrum of an unspiked blank from the sample spectrum (Fig. 11).

Wines are another source of complex phenolic mixtures where phenols may be linked to some desirable sensory characteristics. Figure 12 shows synchronous excitation spectra from a white, red, and a homemade rose wine. (A 20-nm offset was used since three-dimensional scanning demonstrated this offset provides the

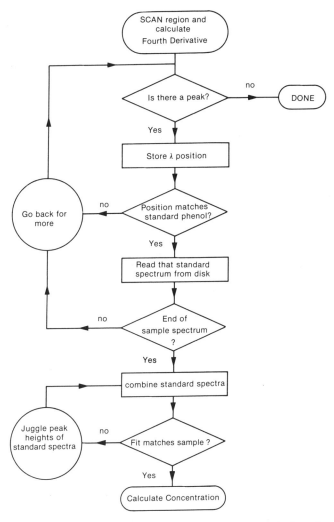

FIG. 8—*Flow chart of a program to locate phenol peaks in a sample through synchronous scanning and the fourth derivative. Once a peak is identified, the concentration of the phenol species responsible for the fluorescence is calculated.*

best signal and component separation.) The red wine was diluted by a factor of 10 to avoid alteration of the signal by self-absorption, so the intensity of the red wine should be an order of magnitude greater than it appears in Fig. 12. The wavelengths of the bands in the wine spectra are higher than those of the common phenols, which is consistent with the fact that wines contain polyphenolic species [*17*]. The peak maxima of the more complex phenolic compounds listed in Table 1 also appear at longer wavelengths.

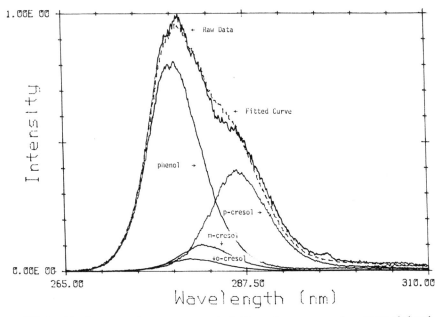

FIG. 9—*The Datamate microprocessor resolved this synchronous scan of a mixture of phenol, o-cresol, m-cresol, and p-cresol into the components shown below. The sum of components* (dashed line) *closely approximates the raw data.*

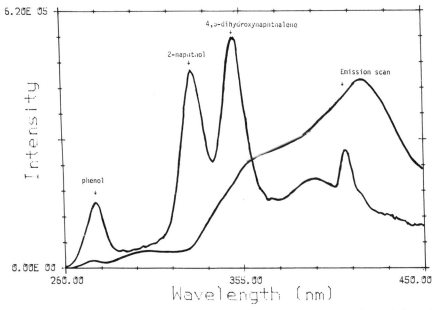

FIG. 10—*A sample from the Raritan Bay was spiked with phenol, 2-napthol, and 4,5-dihydroxynaphthalene. The peaks of the spiked components are resolved by a synchronous scan with a 10-nm offset, though they are unrecognizable in the emission scan.*

PURCELL ET AL ON SYNCHRONOUS EXCITATION 93

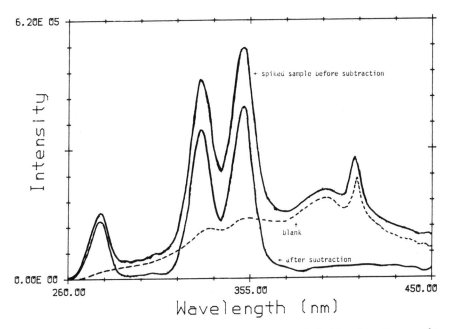

FIG. 11—*Further improvement of a synchronous spectrum can be had by subtracting away the spectrum of an unspiked blank.*

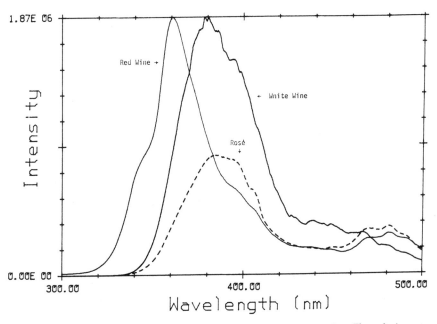

FIG. 12—*Synchronous spectra of three wines, each run with a 20-nm offset. The red wine was diluted by a factor of 10.*

The intensity pattern of the wine spectra correlates well with the known phenolic content of wines in that red wines have the highest concentration of phenols [*17*]. This feature is of special interest since those wines with the highest concentration of phenolic species age the best [*17*].

Conclusions

Synchronous-excitation fluorescence combined with derivative spectroscopy can simply and quickly separate, identify, and quantify mixtures of phenols under the variety of nonideal conditions presented by real-world samples [*10*]. Ease of sample preparation and the speed of analysis are, of course, bonuses. This paper extends the techniques of synchronous scanning by incorporating the fourth derivative and demonstrates on-line separation of four phenolic species. Still, there are other potential improvements that may increase selectivity and extend detection limits. For example, scanning with constant-energy offsets, instead of constant wavelength, is currently under investigation and should make it possible to obtain data with better separation. Also mechanics of energy transfer noticed in these studies merit further attention, since it may prove possible to enhance the spectrum of one phenol species at the expense of another by spiking the sample or altering the offset.

References

[*1*] Boutwell, R. K. and Bosch, D. K., *Cancer Research,* Vol. 19, 1959, pp. 413–424.
[*2*] Boch, F. G. and Burns, R., *Journal of the National Cancer Institute,* Vol. 30, 1963, pp. 393–397.
[*3*] Stofen, D., *Toxicology,* Vol. 1, 1973, p. 187.
[*4*] Birge, W. J., Black, J. A., Hudson, J. E., and Bruser, D. M., *Aquatic Toxicology, STP 667,* American Society for Testing and Materials, Philadelphia, 1979, pp. 131–147.
[*5*] Jerina, D. M., Sayer, J. M., and Yagi, H., *Journal of the American Chemical Society,* Vol. 104, 1982, pp. 5562.
[*6*] Wood, A. W. and Conney, A. H., *Proceedings of the National Academy of Sciences,* Vol. 74, 1982, p. 5122.
[*7*] Salomonsson, A., Theander, O., and Aman, P., *Journal of Agricultural Food Chemistry,* Vol. 26, 1976, p. 830.
[*8*] Gupta, R. N., Eng, F., and Keane, P. M., *Journal of Chromatography,* Vol. 143, 1977, p. 112.
[*9*] Ogan, K. and Katz, E., *Analytical Chemistry,* Vol. 53, 1981, p. 160.
[*10*] Vo-Dinh, T., Gammage, R. B., Hawthorne, A. R., and Thorngate, J. H., *Trace Organic Analysis: A New Frontier in Analytical Chemistry, Proceedings of the 9th Materials Research Symposium,* NBS Special Publication 519, National Bureau of Standards, Washington, DC, 1979, p. 679.
[*11*] Lloyd, J. B. F., *Nature,* Vol. 231, 1971, p. 54.
[*12*] Vo-Dinh, T., *Analytical Chemistry,* Vol. 50, 1978, p. 396.
[*13*] Boto, K. G. and Bunt, J. S., *Analytical Chemistry,* Vol. 50, 1978, p. 392.
[*14*] Vo-Dinh, T. and Gammage, R. B., *Analytica Chimica Acta,* Vol. 107, 1979, p. 261.
[*15*] Cahill, J. E. and Padera, F. G., *American Laboratories,* Vol. 12, 1980, p. 101.
[*16*] Savitsky, A. and Golay, M., *Analytical Chemistry,* Vol. 36, 1967, p. 1627.
[*17*] Amerine, M. A. and Roessler, E. B., *Wines their Sensory Evaluation,* W. H. Freeman and Company, San Francisco, CA, 1976.

Gene Sogliero,[1] DeLyle Eastwood,[2] and James Gilbert[3]

A Concise Feature Set for the Pattern Recognition of Low-Temperature Luminescence Spectra of Hazardous Chemicals

REFERENCE: Sogliero, G., Eastwood, D., and Gilbert, J., "**A Concise Feature Set for the Pattern Recognition of Low-Temperature Luminescence Spectra of Hazardous Chemicals**," *Advances in Luminescence Spectroscopy, ASTM STP 863*, L. J. Cline Love and D. Eastwood, Eds., American Society for Testing and Materials, Philadelphia, 1985, pp. 95–115.

ABSTRACT: As computer libraries of corrected, digitized, low-temperature luminescence, and fluorescence spectra are expanded, it becomes increasingly important to specify a succinct and pithy set of features that will accelerate the library search for a matched spectrum to an unknown sample spectrum. While feature sets for other types of spectra have been investigated extensively, feature sets for fluorescence and luminescence spectra have not been fully explored as yet. Using a specially generated library of low-temperature luminescence spectra of approximately 60 hazardous chemicals, a feature set consisting of only six components (the first four noncentral sample moments of the spectrum, the approximate normalized area under the spectral envelope, and the wavelength corresponding to the location of the maximum intensity) performs exceptionally well in a test using a cluster analysis involving over 2000 pairwise comparisons of the feature sets.

KEY WORDS: luminescence, fluorescence, pattern recognition, low temperature, hazardous chemicals, feature set, spectral moments, cluster analysis, pairwise comparisons, spectral matching, spectral identification

In a previous paper [1], the criteria and instrumental conditions for establishing a computer library of digitized room-temperature fluorescence (RTF) or low-temperature luminescence (LTL) spectra of hazardous chemicals were addressed.

---
[1] Senior research scientist, Department of Clinical Research, Pfizer Central Research, Groton, CT 06340.
[2] Senior chemist, Army Corps of Engineers, Missouri River Division, Omaha, NE 68102.
[3] Technician, Department of Environmental Protection, Fisheries and Wildlife, Hartford, CT 06108.

In the present paper, feature sets suitable for a library of LTL spectra of hazardous chemicals are considered. Although the immediate application is to LTL spectra, the same feature sets and general approach are applicable to RTF spectra as well. The purpose of selecting appropriate feature sets rather than treating each spectrum of nondigitized points as an n-dimensional vector is to reduce the amount of computer storage space required and also to emphasize the characteristic features.

Feature sets [2,3] have been long studied and developed for other types of spectral data, such as infrared (IR) spectra, in order to permit pattern recognition, library searches, and qualitative identification of unknown chemicals. For some types of spectra, such as IR, large computer libraries of many thousands of spectra have been generated. For IR spectra that possess a relatively large number of sharp peaks, a few peak positions (in cm^{-1}) and relative peak heights, or weighting of only the stronger peaks, have been found to suffice for most spectral searches although more elaborate feature sets have also been considered.[4]

Until recently, commercial instruments for generating rapidly completely spectrally corrected and digitized RTF and LTL spectra were not readily available. Also, general agreement on quantitative criteria for evaluating spectral quality for fluorescence and luminescence spectra were also lacking. In most cases, relatively large spectral libraries generated under favorable instrumental and experimental conditions for testing such criteria were also unavailable. For fluorescence spectra, however, only a few studies of feature sets, Lyons et al [4], Yim et al [5], Miller and Faulkner [6], and Mulkerrin and Wampler [7,8]) have been carried out. To the best of our knowledge, the present paper is the first study of feature sets for LTL spectra of hazardous chemicals. For LTL and RTF spectra, because of the relatively small number of peaks and broader peak structure as compared to infrared or laser Raman spectra, totally different feature sets need to be considered including features that indicate the shape distribution of the spectral peaks. Initially, several feature sets were to be considered and compared. The feature sets contemplated for analysis were (1) a histogram of the spectral density (with intervals for grouping the intensities chosen so as to retain the shape characteristics of the spectrum), (2) the cumulative distribution of the spectral density (based on the intervals chosen for the histogram), (3) the locations of the ten highest recorded intensities, and (4) coefficients of the spectrum's Fourier transform [5], plus the one chosen and discussed in this paper. Government cutbacks and time constraints prevented the completion of this goal and resulted in the generation of only 60 to 70 LTL spectra of hazardous chemicals, where initially a library of 200 was envisaged (90 RTF spectra were generated in a separate study [9] but were not included in this analysis). As a result of the government restrictions, only the feature set evaluation, considered to be most promising, was carried out.

It will be shown that the feature set (to be discussed in this paper), which consists of the first four noncentral moments of the spectra, plus two additional

[4]Craver, D., private communication.

features has been successfully applied to a library of LTL spectra. Previous feature sets for fluorescence spectra have been applied only to much smaller sets.

Instrumentation

Low-temperature luminescence spectra were generated by a Farrand Mark I spectrofluorometer, with a corrected spectra accessory, modified to have double excitation monochromators to reduce scatter. A fluorescence free Suprasil® quartz Dewar flask and 3-mm internal diameter sample tube were used with a focusing lens to reduce scatter. Excitation bandwidths were 5 nm and emission bandwidths were 1 nm for all spectra.

Digitized spectra were recorded using a Bascom-Turner Model 810-CRI microprocessor with floppy disk storage. Data were transferred to the Data-General Nova Model 840 computer for processing (to produce the desired spectral signal, normalized and with solvent blank subtracted, and the calculation of the feature sets). Cluster analysis was performed on a DEC-10 computer.

Experimental Procedure

Low-temperature luminescence spectra (at liquid nitrogen temperatures equal to 77 K) for 54 hazardous chemicals were generated for the purpose of examining feature sets that would facilitate the rapid retrieval of matching spectra from a library of digitized spectra. Each spectrum is defined by a set of 500 intensity readings recorded over a spectral range of 250 to 650 nm. The instrument parameters of slit width, damping, and scan speed were set (for the purpose of standardizing the library) at 1-nm emission, 5-nm excitation monochromator slit widths, medium damping, and 30 nm/min respectively. Excitation wavelengths were selected as appropriate to the chemicals in question in order to give the maximum luminescence intensity (as determined by preliminary experiments if not known from the literature or from absorption data). Filters were used when second order scatter interfered with phosphorescence, and also the excitation shutter was closed when scanning through the excitation wavelength. Spectroquality solvents that form clear rigid organic glasses at liquid nitrogen temperature were used. Methylcyclohexane (MCH) was most commonly used as solvent, while a diethyl ether, pentane, and ethyl alcohol (EPA) mixture was used as an alternate solvent when the chemical was insufficiently soluble in MCH. All compounds were prepared at 100 $\mu g/mL$ in the solvent. This concentration was used for all samples partly to simplify sample preparation and partly to ensure sufficient signal from chemicals with relatively low quantum yields. Although at these concentrations some self-absorption and consequent distortion of spectral shape may have occurred for some samples, this would not invalidate the use of our data for illustrative purposes. Just before the generation of the solution spectrum for the chemical, a single spectrum of the solvent blank was obtained and stored.

Procedure for Data Processing

The digitized spectra were transferred to the computer for processing. Special software was written to allow "corrected" spectra to be generated by subtracting solvent blanks; removing artifacts and second order scatter; and renormalization of the spectrum. The Farrand spectrofluorometer contains an instrumental correction to give pure relative emission intensity as a function of wavelength, and this had been checked by comparison with anthracene and other standards for which the corrected luminescence spectra are known.

In addition to the corrected spectra of the 54 compounds, multiple low-temperature luminescence spectra of anthracene, pyrene, and fluoranthene (obtained under eight different experimental conditions) were available for feature analysis. Details of the experimental design and instrument conditions used to produce these additional spectra are described in Ref *1*. This earlier work studied the effects that three instrument parameters, slit width, damping, and scanning speed, have on the spectral shape. Each parameter was analyzed at two levels; a total of eight different treatment combinations were examined. Each treatment combination was replicated, so a total of 16 spectra (for anthracene, pyrene, and fluoranthene) were available for assessing the effectiveness of the features. An analysis of variance (ANOVA) on these three prototype compounds produced a measure of the variability of the feature set between spectra generated under the same and different instrument conditions.

Features were computed for all of the spectra (the technique is described in the next section). Cluster analysis was done on the 16 anthracene, 16 pyrene, and 16 fluoranthene spectra separately, to establish the most clearly matched spectra based on the "features." A cluster analysis was then carried out on the feature sets of the whole library of spectra.

Finally, a cluster analysis was performed on the feature sets of the whole library plus five pairs of the most closely matched sets from the anthracene, pyrene, and fluoranthene groups.

These clusters were examined to assess the selectivity of the features and to establish a threshold.

Description of the Feature Set

The feature set proposed here consists of only six components. These components are

(1) an estimate of the approximate normalized area under the curve (envelope) of the spectral intensities,

(2) the first four noncentral sample moments for the spectrum of each chemical,

(3) the wavelength of the maximum intensity recorded over the range under examination.

Moments were selected because they have always been important in statistics as a means of succinctly characterizing the shape of a distribution function (see Mulkerrin and Wampler for additional information on the analysis of moments of fluorescence spectra [8,9]). Area under the curve was selected because of its importance in discriminating nuances in spectral shape as disclosed in Ref 1. The wavelength of the peak maximum was included because of its intuitive appeal of its importance to spectral shape and its widespread use in infrared pattern recognition studies.

These features can be easily calculated from the stored spectral signal by noting the relationship that exists between the AREA and the moments. Recall that AREA as defined in Ref 1 is

$$\text{AREA} = \sum_{j=1}^{N} S_j \Delta\lambda \tag{1}$$

where $\Delta\lambda$ is the sampling interval wavelength, S_j is the normalized signal intensity at the jth point, and N is the number of points sampled ($N = 500$ for all the spectra produced in the library). The sample moments can be defined as

$$MU(I) = \sum_{j=1}^{N} j^I p_j, \quad I = 0, \ldots, 4 \tag{2}$$

Since $\Delta\lambda$ and AREA are constants (with respect to the summation on j), we can express (1) as $\sum_{j=1}^{N} S_j = 1$ where $K = \Delta\lambda/\text{AREA}$.

Since $MU(0)$ is by definition $\sum_{j=1}^{N} p_j = 1$, then

$$K \sum_{j=1}^{N} S_j = \sum_{j=1}^{N} p_j \tag{3}$$

or

$$\sum_{j=1}^{N} (KS_j - p_j) = 0 \tag{4}$$

Since K, S, and p are all ≥ 0, then

$$KS_j = p_j \quad \text{for} \quad j = 1, 2, \ldots N \tag{5}$$

Equation 5 shows that the normalized signal intensity S_j is related to the density p_j. Thus the moments, as defined by Eq 2, are easily computed from the stored spectral signal, once the AREA has been computed. A simple multiplication of S_j by $K = \Delta\lambda/\text{AREA}$ converts the spectral signal S to a discrete density function of p.[5]

---

[5] Note: p is similar to a "discrete probability density function" which requires that $\sum_{j=1}^{N} p_j = 1$ and that $0 \leq p_j \leq 1$ for all j.

The sixth component of the feature set, the location of the maximum peak, is easily obtained by a simple computer algorithm which searches for the j that corresponds to the maximum S_j. (A five point smoothing done on each spectrum revealed that the highly structured spectra frequently showed a shift in the location of maximum peak, so no smoothing was done on the spectra used in the analysis.)

Cluster Analysis

Automated searches for spectral matches require the calculation of a "distance measure" between the two spectra being compared. The distance measure is often called a "similarity" measure in the literature, in that it quantifies the closeness (or remoteness) of one vector to another. Of the several available measures of similarity, the correlation coefficient and the Euclidean distance are the most often used. For convenience, the Euclidean distance is the similarity measure that is used in this paper. It is chosen because it is needed for the cluster analysis calculation that is used here to test the validity of the feature sets described. In any case, to pick out matched spectra from a computer processing of the feature sets, a similarity measure is required. If the features reflect the important characteristics of the spectral shape, then a small distance will separate the features of the matched spectra. All the matched spectra should have a distance less than some established threshold and all pairings that are below or within the threshold can be selected for closer examination.

The cluster analysis was effected by the BMDP2M[6] program, which is designed to cluster the spectra (or the feature set representation of them). To eliminate the problem of different units for the components, each feature component is standardized. The standardization involves computing the mean and standard deviation of each feature component, and then transforming these components by subtracting the mean and dividing by the standard deviation. This transformation produces a dimensionless number, which replaces the raw calculation of the features. The cluster program then computes the Euclidean distance pairwise between all feature sets in the library. (The number of pairwise comparisons for a library with 63 spectra is 1953, when 10 more are added to the set, the number of comparison becomes 2628.)

The clustering is done by a "single linkage" technique. Initially, each feature set is considered a cluster of one. Feature sets (or clusters of feature sets) are linked in a stepwise fashion, until all cases are part of one cluster. The algorithm for joining clusters requires that the distance between the "centroid of a cluster" and any other cluster, be the shortest distance available at that step. Each time a linkage occurs, the centroid of features of the cluster linked becomes the feature set, at that stage. All pairwise distances are recomputed, the shortest distance is selected, a linkage is made, and a new evaluation is performed. A matrix of the

---

[6]Part of the 1981 BMDP Statistical Software series is from the Department of Biomathematics, UCLA, Los Angeles, CA.

distances between the feature sets before the clustering begins is part of the output that can be requested.

Results

Table 1 gives the code number, name, range, and excitation wavelengths for each of 63 spectra contained in the library developed for this study. Fifty-four different compounds are represented. Several have two or more entries for different wavelength ranges. These repetitions of the same compound were allowed for cases that hinted at some interesting spectral structure in a certain region, but where the structure was not clear because of stronger peaks in a different part of the range. By restricting the spectrum to a limited range and normalizing the intensities for that region, the weaker structure is highlighted.

Table 2 displays the calculated features of the 63 spectra. The code number identifies the compound name via Table 1.

Tables 3, 4, and 5 contain the raw feature sets for anthracene, pyrene, and fluoranthene, respectively. The numbers in the first column correspond to eight different treatment combinations plus replicates. The experimental conditions corresponding to the numbers in Column 1 are shown in Table 6.

For the cluster analysis, the standardized feature sets for anthracene, pyrene, and fluoranthene were used. It is important to note that the standardized values change when additional feature sets are added. The raw feature sets are invariant to the number of feature sets being examined. However, the standardized values are used because they eliminate the problem of dimensions and units of measurement. After standardization, all features are dimensionless and they are reasonably spread about zero.

A tree diagram of the linkage obtained from a cluster analysis of the standardized anthracene feature sets is shown in Fig. 1. The two closest linked spectra (as determined from the features) are (2) and (5). Reference to Table 6 indicates that they were calculated from spectra generated by experimental conditions (1,1,1) = 10 nm/min scan speed, low damping, and 1-nm effective slit width; (1,2,1) = 10 nm/min scan speed, high damping, and 1-nm effective slit width. It is not surprising that replicate spectra were not selected first, since an earlier

TABLE 1 — *Chemicals used in library.*

Code	Compound	Wavelength Range, nm	Excitation Wavelength, nm
1	acenaphthene	310 to 450	290
2	acridine	330 to 506	255
3	azinphos methyl	290 to 590	280
4	azulene	274 to 490	250
5	benzene	266 to 450	250
6	benzoic acid	274 to 530	262
7	benzophenone	274 to 510	260
8	benzyl alcohol	266 to 490	250
9	benzylamine	266 to 490	250
10	bisphenol	382 to 498	270

TABLE 1 Continued.

Code	Compound	Wavelength Range, nm	Excitation Wavelength, nm
11	butyl benzyl phthalate	282 to 538	274
12	carbaryl	290 to 450	285
13	catechol	274 to 410	265
14	1-chloronaphthalene	310 to 450	290
15	p-chlorophenol	282 to 450	280
16	chloropyrifos	300 to 530	280
17	p-chlorotoluene	350 to 590	250
18	p-chlorotoluene	266 to 590	250
19	p-chlorotoluene	266 to 350	250
20	coumaphos	250 to 610	320
21	o-cresol	274 to 490	265
22	2,4-D acid	290 to 550	280
23	dicamba	330 to 570	310
24	dibutyl phthalate	290 to 546	274
25	dichlorobenil	314 to 538	285
26	dichlorobenil	300 to 370	285
27	p-dichlorobenzene	354 to 610	271
28	2,4-dichlorophenol	290 to 610	270
29	2,4-dichlorophenol	290 to 386	270
30	diethylphthalate	290 to 570	270
31	3,5-dimethylphenol	282 to 530	275
32	diphenylamine	290 to 570	290
33	diphenylamine	306 to 386	290
34	diuron	290 to 610	255
35	diuron	274 to 370	255
36	DDD[a]	362 to 530	245
37	DDD	274 to 354	245
38	DDD	374 to 570	245
39	ethylbenzene	266 to 510	255
40	furfural	290 to 610	271
41	furfural	306 to 378	271
42	fluorine	274 to 410	260
43	hydroquinone	290 to 570	290
44	isophthalic acid	290 to 530	280
45	methoxychlor	266 to 530	270
46	naphthalene	290 to 450	265
47	nonylphenol	282 to 530	265
48	pentachlorophenol	290 to 590	250
49	phthalic acid	290 to 530	270
50	pyrogallic acid	274 to 410	270
51	phenylether	282 to 522	265
52	quinoline	410 to 610	277
53	quinoline	306 to 426	277
54	resorcinol	282 to 530	276
55	salicylic acid	330 to 570	300
56	styrene	282 to 410	270
57	tannic acid	290 to 570	276
58	p-toluidine	290 to 546	250
59	toluene sulfonic acid	270 to 490	260
60	2,4,6-trichlorophenol	290 to 650	296
61	tricresylphosphate	274 to 514	260
62	uranyl nitrate	490 to 570	206
63	valeraldehyde	274 to 450	250

[a]DDD is 1,1-dichloro-2,2-bis(p-chlorophenyl)ethane.

TABLE 2—*Feature sets of chemicals.*

Code	Area	MU1	MU2[a]	MU3[b]	MU4[c]	Peak Position, nm
1	25.12	129.84	184.82	291.30	507.00	339
2	61.01	230.13	543.15	1313.25	3249.13	418
3	204.43	235.67	622.20	1784.56	5441.90	457
4	24.28	158.61	267.46	474.88	887.09	376
5	20.66	86.40	114.03	186.44	339.95	279
6	118.30	178.22	371.50	842.26	2015.89	394
7	61.25	245.78	633.81	1675.50	4509.82	445
8	109.78	147.71	271.27	548.93	1176.82	284
9	81.58	138.81	246.34	486.85	1020.43	285
10	74.10	147.92	267.71	544.34	1181.28	303
11	128.69	250.47	661.35	1811.35	5107.07	445
12	33.89	113.00	134.00	16.00	217.00	334
13	42.30	81.28	70.96	66.81	67.99	306
14	28.92	120.78	153.46	207.72	303.39	338
15	38.01	90.22	94.29	120.46	189.99	307
16	119.58	277.66	795.19	2328.56	6941.11	484
17	171.12	273.03	795.49	2453.47	7941.21	436
18	183.89	259.62	745.01	228.37	7393.55	436
19	40.89	66.20	48.22	38.10	32.33	295
20	133.49	251.53	732.08	2360.30	8121.78	359
21	47.37	97.24	132.22	235.88	493.80	297
22	128.55	295.72	909.34	2866.05	9202.03	510
23	105.87	242.24	611.52	1609.74	4417.78	426
24	131.70	249.45	661.69	1827.24	5205.33	446
25	48.50	225.12	524.78	1264.05	3142.25	412
26	40.83	92.37	90.61	94.48	104.37	313
27	171.44	269.74	780.93	2415.23	7920.28	428
28	189.51	292.95	986.28	3526.05	13011.90	538
29	40.64	95.08	93.66	95.66	101.32	319
30	123.96	255.39	681.76	1884.76	5366.23	446
31	58.06	126.60	224.60	495.94	1232.56	298
32	69.50	216.60	501.42	1230.17	3186.76	402
33	49.62	126.69	164.20	217.58	294.44	347
34	178.68	311.76	1017.74	3450.83	12085.10	498
35	43.37	93.35	91.23	93.38	99.97	316
36	156.12	241.61	613.09	1625.59	4476.88	426
37	42.78	73.07	59.88	54.46	53.91	298
38	177.29	250.18	671.15	1903.60	5656.17	426
39	52.65	110.90	182.18	359.76	779.41	277
40	114.37	324.03	1080.16	3685.52	12849.90	490
41	15.72	98.40	103.92	117.16	141.01	319
42	22.94	86.69	83.60	90.22	108.90	303
43	133.52	199.90	447.51	1085.71	2798.29	417
44	86.06	209.68	459.68	1052.48	2514.96	398
45	91.36	186.61	377.52	808.95	1820.69	386
46	23.30	116.76	144.76	192.37	276.33	324
47	77.51	147.49	275.04	589.18	1370.72	301
48	126.78	261.07	710.92	2009.99	5884.72	447
49	122.53	251.63	655.44	1758.56	4844.20	442
50	57.56	81.93	79.89	92.12	122.20	299
51	167.14	201.57	463.41	1143.99	2962.98	417
52	53.32	27.44	1093.70	3727.23	12958.30	490
53	58.07	129.25	177.27	256.62	389.23	330
54	57.48	121.70	196.44	388.16	872.05	295
55	107.69	247.05	633.30	1683.27	4634.58	430

TABLE 2 Continued.

Code	Area	MU1	MU2^a	MU3^b	MU4^c	Peak Position, nm
56	24.73	78.43	66.02	60.21	59.97	294
57	135.23	219.13	558.04	1570.37	4702.89	374
58	151.75	197.72	432.55	1024.03	2578.47	422
59	107.21	167.68	323.03	665.69	1433.67	382
60	171.12	354.42	1298.07	4894.43	18944.50	517
61	56.90	118.36	203.85	482.63	959.95	284
62	37.22	338.48	1151.10	3933.35	13505.40	517
63	70.36	115.16	149.09	208.09	305.91	308

^{a}Quantity in column has been multiplied by 10^{-2}.
^{b}Quantity in column has been multiplied by 10^{-4}.
^{c}Quantity in column has been multiplied by 10^{-6}.

TABLE 3—*Feature sets for anthracene obtained from spectra produced by different experimental conditions.*

Code	Area	MU1	MU2^a	MU3^b	MU4^c	Peak Position, nm
1	27.48	229.97	544.35	1327.5	3335.7	381
2	26.16	229.46	540.30	1306.6	3245.7	381
3	47.00	225.24	518.40	1219.3	2929.3	381
4	52.08	228.69	537.93	1302.7	3248.3	381
5	27.31	229.33	540.42	1309.6	3264.4	381
6	27.57	228.96	538.23	1300.1	3228.0	381
7	51.28	232.01	553.59	1359.5	3436.8	382
8	50.69	228.36	536.08	1294.7	3217.0	382
9	29.69	231.73	551.54	1349.4	3394.1	382
10	30.43	232.71	557.73	1377.8	3508.7	382
11	51.35	230.27	544.45	1322.7	3301.9	382
12	51.89	229.72	542.72	1319.7	3303.6	382
13	33.92	231.53	550.95	1348.4	3394.4	382
14	32.78	220.09	534.88	1285.6	3180.1	381
15	55.18	231.26	549.66	1343.6	3377.7	403
16	54.86	232.62	556.14	1367.4	3457.5	403

^{a}Quantity in column has been multiplied by 10^{-2}.
^{b}Quantity in column has been multiplied by 10^{-4}.
^{c}Quantity in column has been multiplied by 10^{-6}.

study [*1*] established that there was considerable variation in replicate spectra, and also that damping did not significantly influence the shape of the spectrum. (Damping is represented by the second number of the triplet of numbers indicating the experimental factors.) Visual inspection of these two "closest linked" spectra affirms that they are nearly identical. A second linkage is made between (11) and (12). These are replicate spectra, that is, generated under the same instrumental conditions. Figures 2 and 3 show representations of the anthracene spectra to two sets of linked features. Figure 2 corresponds to experimental condition (1,1,1). Figure 3 corresponds to experimental condition (2,1,2).

Figure 4 shows the tree diagram of the cluster linkage of the 16 pyrene feature sets. The two first links are formed from feature Sets 4 and 8 and 9 and 10. Sets

TABLE 4—*Feature sets for pyrene obtained from spectra produced by different experimental conditions.*

Code	Area	MU1	MU2^a	MU3^b	MU4^c	Peak Position, nm
1	54.15	171.28	317.82	639.43	1390.2	383
2	54.37	171.81	318.79	638.78	1376.8	383
3	95.04	171.69	318.88	641.41	1392.9	392
4	93.93	172.38	321.10	646.33	1401.9	392
5	58.45	172.97	323.12	652.09	1416.6	383
6	55.93	172.30	321.95	652.62	1429.4	383
7	95.35	173.13	324.79	660.56	1450.7	392
8	95.54	172.39	321.08	646.49	1402.7	392
9	63.22	173.90	326.40	661.62	1443.7	384
10	62.57	173.41	325.22	660.02	1444.1	384
11	98.00	175.39	331.86	677.56	1487.7	394
12	101.17	177.46	341.11	710.19	1594.5	394
13	69.32	170.03	306.03	581.05	1156.5	385
14	77.76	176.83	336.88	691.74	1525.9	385
15	99.07	176.66	336.26	689.90	1520.6	394
16	100.34	178.04	342.51	712.05	1593.4	394

^{a}Quantity in column has been multiplied by 10^{-2}.
^{b}Quantity in column has been multiplied by 10^{-4}.
^{c}Quantity in column has been multiplied by 10^{-6}.

TABLE 5—*Feature sets for fluoranthene obtained from spectra produced by different experimental conditions.*

Code	Area	MU1	MU2^a	MU3^b	MU4^c	Peak Position, nm
1	220.84	224.20	617.98	1960.5	6873.5	464
2	222.43	211.39	545.13	1602.0	5168.7	464
3	226.89	222.88	603.37	1868.2	6371.1	464
4	216.28	215.47	556.59	1628.5	5211.9	464
5	258.99	248.22	757.16	2627.9	9923.7	464
6	229.34	228.90	641.75	2062.1	7289.4	464
7	225.32	223.41	602.83	1859.2	6318.7	465
8	229.93	224.10	615.52	1942.7	6770.0	463
9	197.76	202.17	472.19	1198.0	3152.0	465
10	222.02	228.77	635.40	2021.0	7084.3	465
11	234.49	230.49	646.89	2077.4	7343.6	466
12	227.26	223.13	602.32	1851.9	6259.4	465
13	233.80	235.88	675.53	2211.6	2957.7	466
14	226.14	229.26	636.40	2015.9	2023.4	438
15	229.36	227.45	622.86	1942.1	6656.9	466
16	202.60	200.27	457.78	1113.2	2707.3	467

^{a}Quantity in column has been multiplied by 10^{-2}.
^{b}Quantity in column has been multiplied by 10^{-4}.
^{c}Quantity in column has been multiplied by 10^{-6}.

4 and 8 were calculated from spectra generated by instrumental conditions (1,1,2) and (1,2,2). A representation of this spectrum is shown in Fig 5. The broad shape of the peaks is due mainly to the 5-nm effective emission slit width as the most important factor in controlling the shape of the spectrum. Feature Sets 9 and 10

TABLE 6 — *The experimental conditions corresponding to the numbers in Column 1 of Tables 3 through 5.*

Code	Experimental condition[a]			Speed, nm/min	Damping	Slit, nm
1	1	1	1	10	low	1
2	1	1	1	10	low	1
3	1	1	2	10	low	5
4	1	1	2	10	low	5
5	1	2	1	10	high	1
6	1	2	1	10	high	1
7	1	2	2	10	high	5
8	1	2	2	10	high	5
9	2	1	1	60	low	1
10	2	1	1	60	low	1
11	2	1	2	60	low	5
12	2	1	2	60	low	5
13	2	2	1	60	high	1
14	2	2	1	60	high	1
15	2	2	2	60	high	5
16	2	2	2	60	high	5

[a]The treatment combinations are represented by the triplet of 1s and 2s shown in the column for experimental condition. 1 refers to the low level of the factors, and 2 refers to the high level. Low level of the three factors is 10-nm/min scan speed, low damping and 1-nm effective emission slitwidth. High level of these factors is 60-nm/min scan speed, high damping, and 5-nm effective emission slitwidth.

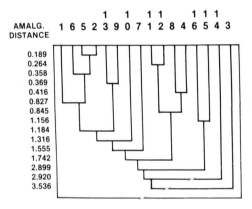

FIG. 1 — *Tree diagram of cluster linkages for anthracene feature sets.*

come from calculations of spectra generated by the experimental condition (2,1,1). Figure 6 shows a representative spectrum for this condition.

Figure 7 shows the tree diagram of the linkages made from a cluster analysis of the fluoranthene feature sets. As discussed in Ref *1*, fluoranthene has a broad spectrum with no sharp peaks. Its spectral shape was not significantly altered by the experimental conditions used so there was little discernible variation in the spectrum. The main variability with fluoranthene is caused by random fluctu-

SOGLIERO ET AL ON PATTERN RECOGNITON 107

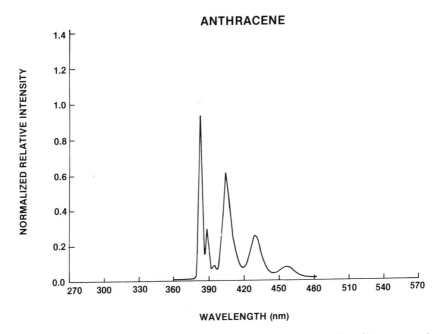

FIG. 2—*Spectrum of anthracene—Experimental conditions (1,1,1) are 10-nm/min scan speed, low damping, and 1-nm effective emission slitwidth.*

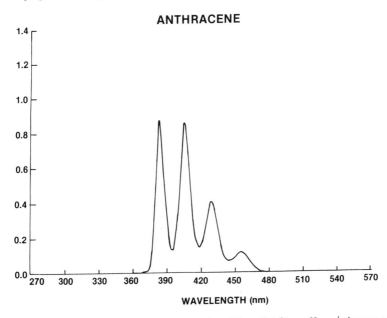

FIG. 3—*Spectrum of anthracene—Experimental conditions (2,1,2) are 60-nm/min scan speed, low damping, and 5-nm effective emission slitwidth.*

108 ADVANCES IN LUMINESCENCE SPECTROSCOPY

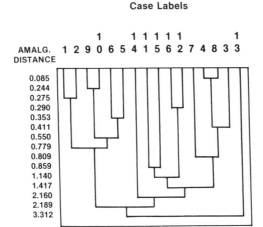

FIG. 4—*Tree diagram of cluster linkages for pyrene feature sets.*

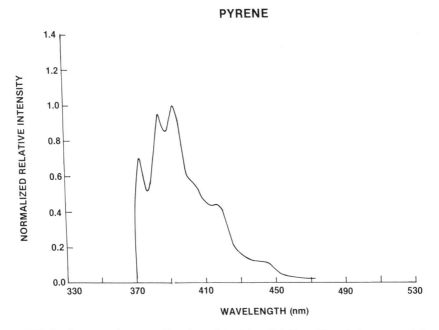

FIG. 5—*Spectrum of pyrene—Experimental conditions (1,1,2) are 10-nm/min scan speed, low damping, and 5-nm effective emission slitwidth.*

ation. Figure 8 shows a spectrum representative of the experimental conditions (1,2,2) and (2,1,2), which are the closest linked Sets 7 and 12.

Figure 9 shows the tree diagram for the set of features for the 63 spectra in the library (Table 2). The spectral code labels are displayed across the top, the

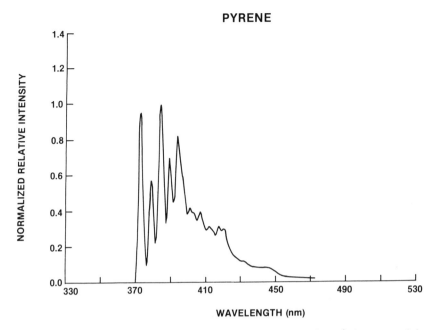

FIG. 6—*Spectrum of pyrene*—*Experimental conditions (2,1,1) are 60-nm/min scan speed, low damping, and 1-nm effective emission slitwidth.*

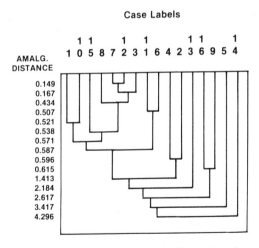

FIG. 7—*Tree diagram of cluster linkages for fluoranthene feature sets.*

amalgamation distances on the side. Two of the early linkages are between butyl benzyl phthalate (11) and dibutyl phthlate (24), and bisphenol (10) and non-ylphenol (47). Figures 10 and 11, and Figs. 12 and 13 show the spectrum of these compounds. The similarity is evident. Since there were no replicate spectra of

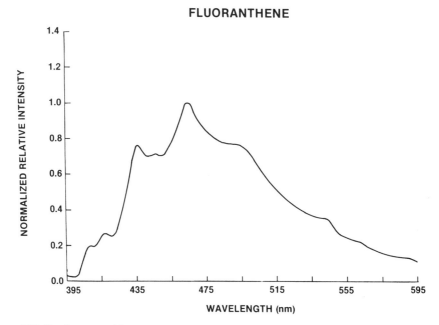

FIG. 8 — *Spectrum of fluoranthene* — *Experimental conditions (2,1,2) are 60-nm/min scan speed, low damping, and 5-nm effective emission slitwidth.*

of any of the compounds represented in the library, the selections made by the cluster algorithm from the features are quite good.

To test the validity and efficacy of the feature sets, ten additional spectral feature sets were added to the library. The sets corresponded to the five matched sets from the anthracene, pyrene, and fluoranthene diagrams, the ones represented by Figs. 2, 3, 4, 5, and 8 as discussed earlier. These ten additional feature sets, brought the total of spectra represented in the library to 73. This means that 2628 distinct calculations corresponding to all pairwise comparisons had to be made. The question to be answered was whether or not the features would be good enough to select the proper matches.

The cluster analysis for the 73 spectra is displayed by the tree diagram of Fig. 14. Inspection of the first five linkages show them to be

(1) Codes 67 and 66 — anthracene pairs (2) and (5),
(2) Codes 70 and 71 — pyrene pairs (4) and (8),
(3) Codes 64 and 65 — anthracene pairs (11) and (12),
(4) Codes 68 and 69 — pyrene pairs (9) and (10), and
(5) Codes 72 and 73 — fluoranthene pairs (7) and (12).

All five appropriately matched spectra were selected first and all had an amalgamation distance <0.040. A look at the next link comes at a distance of

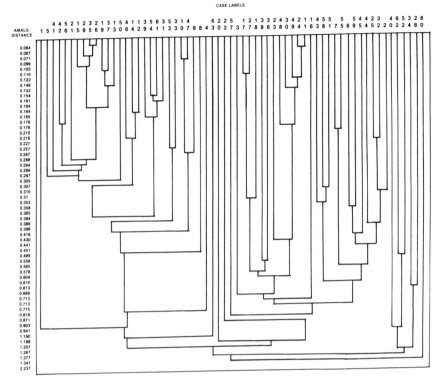

FIG. 9—*Tree diagram of cluster linkages for 63 feature sets.*

0.064 for the spectra (11) and (24) shown in Figs. 12 and 13. A conservative threshold of ≤0.05 might be used for this library in order to prevent an undue number of false alarms and yet allow for true matches.

A question might also be asked about the level at which the two pairs of non-linked anthracene and pyrene spectra might be joined. Although they are representations of the same compound, their spectra are manifestly different (Figs. 2 and 3, and 5 and 6). They do not join until much later. This illustrates the importance of standardization of the instrument conditions and following the procedure of generating the library and testing samples. Otherwise, no features (not even the full spectra of all the intensities) will work.

A histogram of the 2628 pairwise distances between the 73 feature sets shows that 23 feature sets have a distance of ≤0.188, while the five matched spectra had a distance of ≤0.05, which would be a good threshold to use for this library. For example, if a luminescence spectrum of an unknown compound (taken from the chemicals in the spectral library) was generated by the procedure described in this report, the distance for the match with the appropriate spectra taken from the library would be less than 0.05 most of the time.

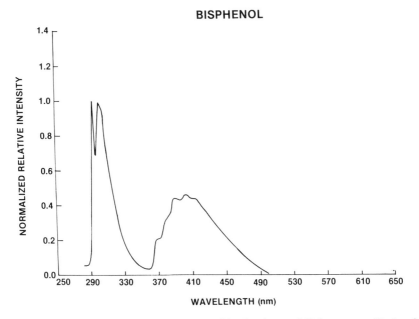

FIG. 10—*Closely linked spectra as determined by the cluster of 63 feature sets. Bisphenol— Code 10.*

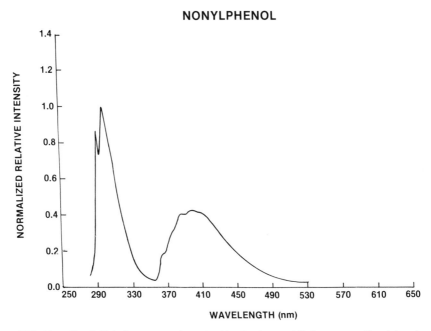

FIG. 11—*Closely linked spectra as determined by the cluster of 63 feature sets. Nonylphenol— Code 47.*

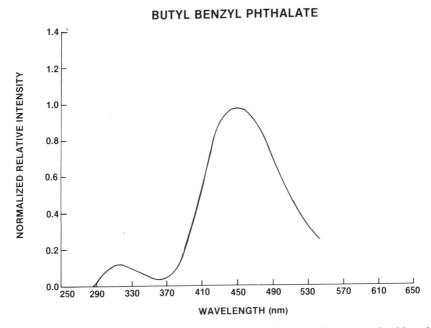

FIG. 12—*Closely linked spectra as determined by the cluster of 63 feature sets. Butyl benzyl phthalate—Code 11.*

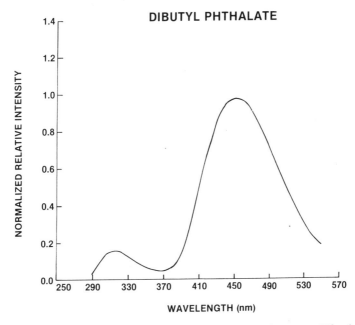

FIG. 13—*Closely linked spectra as determined by the cluster of 63 feature sets. Dibutyl phthalate—Code 24.*

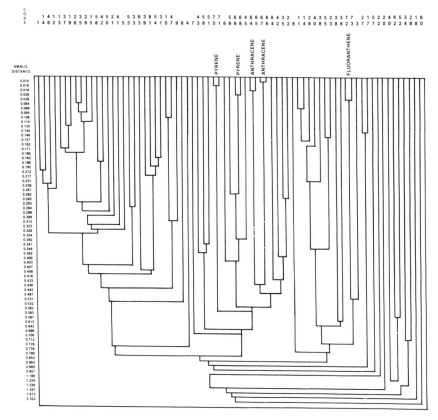

FIG. 14—*Tree diagram of cluster linkages of 73 feature sets.*

Discussion

Although this library of spectra is relatively small, the feature set as defined by the six components appears to do well. Tests were run to ascertain the importance of including the *peak location*. This involved performing the analysis when the peak location information had been excluded. The cluster carried out in this manner showed seven linkages below the threshold of 0.05. Thus, two other sets of spectra were included. Since there was only five matches in all, the additional two were "false alarms." With a larger library, omission of this component would let in many more false matches, and it is therefore deemed important to hold peak position as one of the feature components.

Before establishing the protocol for procuring the spectral signal for the library, a five-point smoothing was done on each spectrum to see the effect of the smoothing on the features. For the most part, the moments were essentially invariant, showing only minor changes. The peak location for broad spectra, such as obtained with fluoranthene, was not affected, and the smoothing removed some of the random noise. For the highly structured spectra (such as anthracene and pyrene) there was, however a shift in the peak position, and often the first and second peaks as ranked by height changed places. As a result of the peak positions' sensitivity to smoothing, no smoothing was performed on the spectral signals stored.

Conclusions

A set of features that are easily calculated from a digitized spectral signal has been found to be sensitive and accurate enough to produce the appropriate matches from a library of spectra. But underlying the success of this or any feature set is the need for a protocol for generating the library. A standardization of the procedure for developing the library, such as the one described here and processing an unknown chemical spectrum, must be addressed at the start if the pattern recognition technique is to be effective.

Although the authors cannot at this time pursue this work further, it is hoped that this approach will be used in the development of a truly extensive library of LTL and RTF spectra, and that the features described herein will be used for rapid search, sample matching, and identification of unknown chemicals.

References

[1] Sogliero, G., Eastwood, D., and Ehmer, R., *Applied Spectroscopy,* Vol. 36, No. 2, 1982, pp. 110–116.
[2] Jurs, P. C. and Isenhour, T. L., *Chemical Applications of Pattern Recognition,* Wiley Interscience, New York, 1975.
[3] Tamara, T., Tanabe, K., Hirashi, J., and Sacki, S., *Bunselic Kagenber,* Vol. 28, 1979, p. 591.
[4] Lyons, J. S., Hardesty, P. J., Baer, E. S., and Faulkner, L. R., *Fluorescence Spectroscopy,* Vol. 3, E. L. Wehry, Ed., Plenum, New York, 1981.
[5] Yim, K. W. K., Miller, T. C., and Faulkner, L. R., *Analytical Chemistry,* Vol. 49, 1977, p. 2009.
[6] Miller, T. C. and Faulkner, L. R., *Analytical Chemistry,* Vol. 48, 1976, p. 2083.
[7] Mulkerrin, M. G. and Wampler, J. E., *Analytical Chemistry,* Vol. 54, 1982, p. 1778.
[8] Mulkerrin, M. G. and Wampler, J. E., *Abstracts of Pittsburgh Conference on Analytical Chemistry and Applied Spectroscopy,* Paper 059, Atlantic City, NJ, March 1982.
[9] Brownrigg, J. T., Busch, D. A., and Giering, L. P., *A Luminescence Survey of Hazardous Materials,* Baird Corporation for U.S. Coast Guard Research and Development Center, Dept. CG-D-53-79 May 1979.

Summary

Three areas of luminescence studies emerge from the selection of papers in this volume. One general area addressed by four of the papers concerns the sensitivity of luminescence phenomena to respond to small molecular perturbations such as the lumiphors' microenvironment. Another area on which two papers focus involves complementary or indirect measurements where competing phenomena such as radiational deactivation, quenching or complexation by the lumiphor provide useful information. The third area reflects the growing use of computer-assisted luminescence measurements where additional spectroscopic information can be extracted by transformation of the spectral data. These areas reflect the continuing sophisticated development of fluorescence and phosphorescence spectroscopy.

The paper by Wirth and coworkers discusses the fundamental nature of the solvation process. These solvation interactions can profoundly change both the spectroscopic characteristics as well as the chromatographic characteristics of molecular species. The difficulty in obtaining a better understanding of liquid theory has been eased by application of two-photon spectroscopy, fluorescence depolarization, and rotational diffusion studies by the authors to identify spatial properties. Using sophisticated instrumental approaches, the authors have correlated the effects of molecular shape on solvation interactions using both polar and nonpolar solvents and polycyclic aromatic hydrocarbon lumiphors. It is postulated that solvent structures composed of molecules immediately surrounding a solute molecule can be shape-selective and respond differently to molecular species based on their structure.

The results of McMorrow and Kasha on the photo-excitation steps in 3-hydroxyflavone, leading to phototautomerization in the ultra-rapid (<8 ps) regime producing an unique tautomer emission, has proven useful in detecting hydrogen-bonding impurities in solvent. Ethers, alcohols and water, and other H-bonding solvents interact with the intramolecular transfer of the hydroxyl hydrogen to the neighboring carbonyl group producing changes in the fluorescence spectrum. The use of 3-hydroxyflavone as a fluorescence probe for solvent impurities is an example of using subtle microenvironmental effects to detect 10^{-7} to 10^{-9} M water contamination.

The study by Chen and Scott concerns the use of fluorescence polarization to follow the dynamic motion of entire fluorescence-tagged proteins (global) or the more rapid motion of particular nonglobal moieties such as a dansyl label. The results show that dansyl conjugates exhibit both a thermally activated rapid rotation and a viscosity independent rotation by steady state analysis, that rapid and slow rotations can be visualized directly by time-resolved anisotropy measurements, that some rotational freedom of subunits is possible in certain cases, and that global rotations can be obtained from steady state isothermal measurements of intrinsic protein polarization. The implication of these rapid fluctuations helps explain, for example, the action of antibodies and enzymes, how proteins undergo transitions between conformational states, and many binding phenomena.

Weinberger et al continue their definitive study of room temperature phosphorescence (RTP) in fluid solution with a discussion of four techniques: micelle-stabilized RTP, microcrystalline/colloidal RTP, cyclodextrin-induced RTP, and sensitized/quenched RTP. The analytical utility of these methods is demonstrated for several classes of chemicals including carbo- and heterocyclic aromatic hydrocarbons and drugs.

Kirkbright's paper develops the theoretical foundation for determination of absolute quantum efficiencies for both solid and liquid samples using the complementary technique of photoacoustic spectroscopy. In this technique, heat at the surface of the sample resulting from nonradiative relaxation of the excited states is measured with respect to wavelength and can give information on the quantum efficiencies of molecular species. The procedure by which this is accomplished with simple instrumentation and without reference to luminescence standard materials is described.

Seitz and coworkers developed a metal ion sensor by immobilizing a fluorogenic ligand on the tip of a bifurcated fiber optic. The ligands successfully used in this fashion included morin, quercetin, and calcein onto cellulose using cyanuric chloride as a coupling reagent.

Purcell et al demonstrate the utility of synchronous excitation fluorescence (SEF) spectroscopy as a highly sensitive and selective technique for the determination of hazardous chemicals, in particular, phenols. In conjunction with derivative spectroscopy, SEF can detect and discriminate among phenols, even isomers, in the parts per billion range.

The paper by Sogliero et al involved generation of a computer library of the low-temperature luminescence spectra of approximately 60 hazardous chemicals. A feature set for each molecular species, namely, a set of only six specific features uniquely characteristic of the compound, were identified to allow rapid matching of an unknown sample spectrum to the stored standard spectrum. The six feature components used were the first four noncentral sample moments of the spectrum, the approximate normalized area uner the spectral envelope, and the wavelength corresponding to the location of the maximum intensity. The feature sets performed very well in a test using a cluster analysis involving over 2000 pairwise comparisons.

The broad scope covered by these papers is an indication of the many diverse fields that use luminescence techniques and of the multitude of applications for

SUMMARY

which luminescence spectroscopy is ideally suited. The editors hope this volume will provide the background and direction to other researchers in the field to develop additional novel uses of light-induced phenomena to form the basis of subsequent publications in this series.

L. J. Cline Love
Seton Hall University, Department of Chemistry, South Orange, NJ, 07079; symposium co-chairman and co-editor.

DeLyle Eastwood
U.S. Army Corps of Engineers, Missouri River Division Laboratory, 420 S. 18th St., Omaha, NE 68102; symposium co-chairman and co-editor.

Index

A
Anisotropy, 12, 26

B
Bandwidth, 9, 35
Beer's law, 57
Biacetyl, 48

C
Calorimetric methods, 56
Chemical microenvironment probes, 5
Chemicals
 Hazardous — methods of detection, 95
Chemiluminescence measurements
 Micellar solutions, 42
Chen, Raymond F., 26–39
Chloride
 Cyanuric, 69
 Ions, 57
Chromatography, 6
 Chromatographic phenomena, 8
 Reverse phase, 9
 Micelle separation, 42
Cline Love, L. J., coeditor, 1, 2, 40–51, 117–119
Cluster analysis — pairwise comparisons, 100
Cyclodextrin (CD-RTP), 49

D
Dansyl conjugates, 26

Rotational freedom
 Table, 34
Dephasing time, 9
Diffraction efficiency, 7
Diffuse reflectance spectroscopy, 62
Dimethylnapthalene isomers, 9
Ditzler, Mauri A., 63–77

E
Eastwood, D., coeditor, 1, 2, 95, 115, 117–119
Electromagnetic radiation, 55

F
Feature sets
 Description, 98
 Library development of chemicals
  Tables, 101, 103, 104, 105, 106
 LTL/RTF spectra — hazardous chemicals, 96
 Noncentral moments, 96
 Wavelength maximum, 98
Florescence
 Decay, 31
 Depolarization, 27
 Emission, 17
 Fiberoptic, 63
 Micelle enhancing, 42
 Polarization, 27
  Relaxation parameters
   Table, 36
 Quantum efficiency utilizing PAS, 56

Quenching, 57, 63
Room temperature (RTF) spectra, 95
Library development of hazardous chemicals, 95
Synchronous-excitation (SEF), 81
Fluorescent complexes, 63
Fluorogenic ligand, 63
Immobilization, 63
  Design — amount/procedure/substrate, 69
  Instrumentation, 70
  Theory, 65
Immobilized ligands
  Calcein, 74
  Hydroxynaphthol blue, 72
  8-Hydroxyquinoline-5-Sulfonate, 74
  Morin, 71
  Quercetin, 71
  Salicylidene-o-Aminophenol, 74
  p-Tosyl-8-Aminoquinoline, 76
  2, 2', 4-Trihydroxyazobenzene, 76

G

Gilbert, James, 95–115

H

Hahn, David, A., 5–15
H-bonding impurities — qualitative discrimination, 19
H-bonding interactions, 18
Hudson, Robert D., 63–77
3-Hydroxyflavone, 17
  Absorption/emission, 18
  Behavior in solvents, 19
  Excitation spectra, 20
  Fluorescence emission spectra, 21
  Fluorescence probe, 22
  Fluorescence spectra, 19

I

Instrumentation, 41, 55, 70, 83, 97
Intramolecular effects, 17
Isolated-site crystal matrix
Shpolskii matrix, 22

K

Kaminski, Raymond, 81–94
Kasha, Michael, 16–25
Kirkbright, Gordon F., 55–62
Koskelo, Aaron C., 5–15

L

Liquid chromatography (LC), 6, 47, 83
Lorentzian, 9
Luminescence
  Coupled phenomena, 55
  Efficiency
    Quantum, 58
    Tables, 61
  Low temperature (LTL) spectra, 95
  Manipulation of observable, 81
  Spectrometry, 56
  Standard materials, 48

M

McMorrow, Dale, 16–25
Methylcyclohexane, 17
Micelle-stabilized (MS-RTP)
  Aromatic compounds, 42
  Heavy atom concentration, 42
  Limits of detection, 44
  Surfactants, 43
  Triplet-triplet annihilation (TTA), 43
Microcrystalline phosphorescence, 44
  RTP wavelengths
    Table, 46
  Colloid similarity, 45
Mohler, Carol E., 5–15

N

Nanosecond spectroscopy, 14
Nonexponential phosphorescence decay, 23
Nonradiative relaxation, 55

O

Obenauf, Ralph H., 81–94

P

Pattern recognition, 96
Perrin plots, 29
Phenols, 81
  Automation, 88
    Illustration, 91
  Benefits, 82
  Detection methods
    Derivative spectroscopy, 87
    Fluorescence, 84
    Synchronous scanning, 85
      Limits of detection, 86
      Tables, 88, 89
      Three dimensional, 86
        Illustration, 87
  Hazards, 82
  Instrumentation — spectrofluorometer, 83
    Illustration, 83
  Occurrence, 82
  Structure
    Illustration, 82
Phosphorescence
  Molecular association, 50
  Room temperature (RTP) in fluid, 40
Photoacoustic
  Signal saturation, 59
  Spectroscopy (PAS), 55
  Studies, 55
Photoautomerization, 18
Photon polarization ratio, 9
  Illustration, 10
Picosecond spectroscopy, 14, 18
Pokornicki, Steven, 63–77
Polycrystalline, 22
Polycyclic aromatic hydrocarbons (PAH), 9, 44
Polynuclear aromatic compounds (PNA), 84
Protein
  Labeled, 27
  Unlabeled, 32
Protein rotations, 27
  Classes
    Differential polarized phase fluorometry, 32
    Steady state, 28
    Time-resolved fluorescence anisotropy, 30
  Global, 29
  Nonglobal, 26
  Segmental, 32
Protein structure
  X-ray crystallography, 26
Proton transfer, 17
  Illustration, 17
Proton transfer spectroscopy — analytical application, 16
Purcell, Francis J., 81–94
Pure liquid structure, 7

Q

Quantum efficiency, 55
Quenched phosphorescence, 40
  Cyclodextrin, 43
  Microcrystals, 43
Quinine bisulfate, 57

R

Radiative relaxation, 55
Rembish, Karin, 40–51
Rotational diffusion constants, 9

S

Saari, Linda A., 63–77
Sanders, Matthew J., 5–15
Scott, Carrie H., 26–39
Seitz, W. Rudolpf, 63
Sensitized RTP, 47
Sensors
  Fluorescence, 64
    Bifurcated fiber optic
      Illustration, 70
  Metal ion, 63
    Illustration, 64
Sieber, Steven C., 63–77
Sogliero, Gene, 95–115
Solute
  Chromophores, 7
  Shape, 12
Solvation interaction, 5
  Dynamics, 7
  Environment, 10
    Molecular shape, 6
  Intermolecular, 6
    Attractive, 7
      Illustration, 8
    Repulsive, 6
      Illustration, 8
    Solvent solute, 12
  Liquid theory, 5
  Spectroscopic, 7
    Raman, 12, 86, 96
    Two-photon, 12
Structure, 6
Spectral data — computer assisted
  Identification, 95
  Matching, 109
  Moments, 99
  Searches, 96
  Structure, 101

T

Tautomerization, 17
  Intramolecular, 16
Tetracene, 12
  Illustration, 13
Trace determinations, 19
Two-photon
  Perturbation calibration — ab initio/empirical, 11
  Spectroscopy, 7
  Symmetries, 11

W

Weinberger, Robert, 40–51
Wirth, Mary J., 5–15

X

Xanthone, 23

Z

Zhujun, Zhang, 63–77

ISBN 0-8031-0412-X